"This is an historical geography of Britain that is both horse-drawn – drawn through the figure of the horse – and horse-powered – showing how the power of equine bodies (to carry, lift, trot, gallop, and more) has shaped the country's changing human landscapes. Whether packhorse or post-horse, carthorse or racehorse, workhorse or war horse, Neil Ward tellingly captures their crucial roles in connecting spaces, quickening human circulation, and making places, spurring innovation in work, infrastructure and social relations. This book is hence that rare thing, one that will intrigue lay readers, engage serious scholars and inspire all who are fascinated by big questions such as those about human-animal co-existence."

—**Chris Philo**, *editor of Animal Spaces, Beastly Places*

"Neil Ward expertly provides a geographical and historical analysis of social power and place-making activities around the lives of horses in Britain. While we read less of individual horse lives and their unique personalities, it is clear how this species' personable nature has created a large horse-shaped hoofprint across contemporary society, and not only in rural areas. Ward's analysis raises critical questions about the consequences of human-horse relations, not only for the quality and value of various horse lives lived, but also their specific contribution to producing social harms and ongoing inequalities."

—**Emma Roe**, *University of Southampton*

"Neil Ward canters through time and space to provide a much-needed study on a national scale, linking fragmented academic fields and providing original observations which bring horse-human relations right up to date. From war horses to therapy horses – and from agricultural labourers to climate-impacting consumers – Ward delivers a thought-provoking, must-read overview for students, scholars and policymakers alike."

—**Tom Almeroth-Williams**, *author of City of Beasts*

"*Horses, Power and Place* is essential reading to anyone interested in horses, 'horsescapes' and animal and equine geographies. It provides a rich and much-needed analysis of the histories and geographies of horses in Britain. Mobilising a more-than-human framework, Ward shows how different human-horse relations manifest over time. This includes the essential role of the horse in agriculture, transport and industry, and in war times, combined with more familiar human-horse relations, notably the racehorse. By emplacing the horse in different historical and contemporary contexts and uses, it stretches where and how we associate horses with place-making. A fascinating book, which I highly recommend – worth a punt, you won't be disappointed!"

—**Damian Maye**, *University of Gloucestershire*

"*Horses, Power and Place* is an intriguing and unique study of a significant intersection of geography and animal studies: the spatialized relationship between humans and horses in Britain and its empire. Ward provides a fascinating exploration of how that relationship has changed over time, as the use of horses transitioned and narrowed from the practical – in transport, agriculture, industry, and war – to their more recent and restricted employment in sporting and leisure activities. This book is a must-read for anyone interested in equine history."

—Allyson N. May, *University of Western Ontario*

"*Horses, Power and Place* is a long-needed historical geography of all things horse. Packed with fascinating historical titbits, the book illuminates the role of horsepower – literal and figurative – in the history of human society. Taking the reader on a ride through horse past and present, Ward reveals the uncertain future of the horse in the face of economic, ethical and environmental challenges. Insightful and entertaining, a must read for academics, students and others interested in how animals have shaped our world."

—Lee-Ann Sutherland, *The James Hutton Institute*

"*Horses, Power and Place* provides a thought-provoking and holistic approach to thinking about the place horses hold in our society, and how they, in turn, have shaped the places we inhabit. Covering wide-ranging phenomena from early equine domestication through to modern relationships with horses, this book has something for everyone with an interest in the role horses have in our lives."

—Tamzin Furtado, *University of Liverpool*

Horses, Power and Place

Horses, Power and Place explores the evolution of humanity's relationship with horses, from early domestication through to the use of the horse as a draught animal, an agricultural, industrial and military asset, and an animal of sport and leisure.

Taking an historical approach, and using Britain as a case study, this is the first book-length exploration of the horse in the more-than-human geography of a nation. It traces the role and implications of horse-based mobility for the evolution of settlement structure, urban morphology and the rural landscape. It maps the growth and various uses of horses to the point of 'peak horse' in the early twentieth century before considering the contemporary place of the horse in twenty-first century economy and society. It assesses the role of the horse in the formation of places within Britain and in the formation of the nation. The book reflects on the implications of this historical and contemporary equine geography for animal geographies and animal studies. It argues for the study of animals in general in how places are made, not just by humans.

Written in a clear and accessible style, this book will be essential reading for students and scholars of animal geography and animal studies more widely.

Neil Ward is Professor of Rural & Regional Development at the University of East Anglia in Norwich, where he was Deputy Vice Chancellor (2014–2021). He was Director of the Centre for Rural Economy at Newcastle University (2004–2008), served as a Cabinet Office advisor on agriculture and rural affairs, and is author of *Net Zero, Food and Farming: Climate Change and the UK Agri-Food System* (Routledge, 2023).

Routledge Human–Animal Studies Series
In Memory of Professor Henry Buller, founding series editor

The new *Routledge Human–Animal Studies Series* offers a much-needed forum for original, innovative and cutting-edge research and analysis to explore human–animal relations across the social sciences and humanities. Titles within the series are empirically and/or theoretically informed and explore a range of dynamic, captivating and highly relevant topics, drawing across the humanities and social sciences in an avowedly interdisciplinary perspective. This series will encourage new theoretical perspectives and highlight ground-breaking research that reflects the dynamism and vibrancy of current animal studies. The series is aimed at upper-level undergraduates, researchers and research students as well as academics and policy-makers across a wide range of social science and humanities disciplines.

The Imaginary of Animals
Annabelle Dufourcq

Winged Worlds
Common Spaces of Avian-Human Lives
Edited by Olga Petri and Michael Guida

Methods in Human-Animal Studies
Engaging with animals through the social sciences
Edited by Annalisa Colombino and Heide K. Bruckner

What We Owe to Nonhuman Animals
The Historical Pretensions of Reason and the Ideal of Felt Kinship
Gary Steiner

Horses, Power and Place
A More-Than-Human Geography of Equine Britain
Neil Ward

For more information about this series, please visit: www.routledge.com/Routledge-Human-Animal-Studies-Series/book-series/RASS

Horses, Power and Place

A More-Than-Human Geography of Equine Britain

Neil Ward

LONDON AND NEW YORK

First published 2024
by Routledge
4 Park Square, Milton Park, Abingdon, Oxon OX14 4RN

and by Routledge
605 Third Avenue, New York, NY 10158

Routledge is an imprint of the Taylor & Francis Group, an informa business

British Library Cataloguing-in-Publication Data
A catalogue record for this book is available from the British Library

ISBN: 978-1-032-59358-6 (hbk)
ISBN: 978-1-032-59359-3 (pbk)
ISBN: 978-1-003-45435-9 (ebk)

DOI: 10.4324/9781003454359

Typeset in Times New Roman
by SPi Technologies India Pvt Ltd (Straive)

Contents

Illustrations

Figures

Tables

Preface

It has been half a century since I sat on a horse. I have not had much to do with the beasts, but, thinking back, they were often there in the background clip-clopping around the West Norfolk village where I grew up. My personal experience with horses can be framed by three separate episodes. First, there were the horses of my early childhood. Walking to school in the early 1970s, I would stop to greet Tarquin, a beautiful chestnut gelding with a white mark down his face who lived in a fenced meadow, with his name on the gate. Tarquin is the only horse I have ever known by name. At around that time, at the seaside, I might sometimes get treated to a ride on a small pony as it shuffled up and down the beach, usually led by a despondent-looking youth. On Saturday afternoons, after the excitement of the lunchtime boxing or tobogganing, televised sport would turn to horseracing from somewhere like Newmarket or Chepstow. I remember this being a deflating moment, and the point at which I would turn off. I could not get excited about horseracing. Despite my rural upbringing, horses were marginal figures, around in the background, but not really of much interest to me.

The second episode comes almost 20 years later in 1990. It was during Margaret Thatcher's last few months in power, although this was not known at the time. She was introducing her 'poll tax' to finance local government, which felt deeply unfair. On a warm spring Saturday, I joined tens of thousands of others who gathered in Kennington Park in South London for a protest march which was to end in a rally in Trafalgar Square. Up until that day, the event was being called 'the Poll Tax March', although it would come to be known by a different name. The march felt quite good natured. The streets rang to the chant of "Maggie, Maggie, Maggie – out, out, out!", and placards proclaiming "Scrap the Tory Poll Tax" were waved merrily. It was a normal, almost routine, political demonstration as we wove our way towards Westminster. However, when I reached Whitehall, I saw people sitting down in the road outside Downing Street, which was hampering the progress of the march. The atmosphere began to change and became menacing. There were riot police on horseback. I took fright, gave up on the idea of listening to the speeches at the rally and caught a bus home to Haringey. There I saw the TV news coverage of the rioting that broke out in and around Trafalgar Square after I had left. Lines of mounted police charged the crowd, and people were trampled under hooves. The scene looked medieval. I was relieved to have escaped the violence and aghast to see horses ridden straight at protestors.

More than 50 years after my encounters with Tarquin, and more than thirty since the Poll Tax Riots, I struggle to think of any other personal experience that has involved horses. Then, a third episode. In 2021 I was looking at the British system of land-based colleges that specialise in agriculture and related industries, and sometimes provide courses in equine science. I was interested in the prospects for institutions in this part of the higher education sector. Where were the growth opportunities and what were the challenges? I found that the equestrian sector accounts for several billions of pounds of annual spending in the UK, has several million regular participants and is a major land user. The size of the sector seemed remarkable, and a set of questions began to nag. What is the nature of the horse economy and how is it changing? Is the British equestrian sector typical or distinctive compared to other countries? Who is responsible and speaks for the horse sector and what are their preoccupations? As I wondered about these contemporary empirical questions, they were joined by more historical and conceptual musings. What, actually, *are* horses and where do they come from? How has the horse shaped our world and its evolution? How is our thinking about the horse changing? In short, what is the historical and contemporary geography of the horse?

Quaint features of country life and seaside resorts, intimidating instruments of state power and an economic sector facing change – my three personal equine encounters reflect three different perspectives on the horse. There are many others. There are almost 60 million horses in the world, including seven million in Europe, a similar number in the United States, and two million in Australia. Horseracing is a global industry that involves millions of followers through betting and as spectators. Worldwide, most horses are still working animals, but in the Western world they are now more likely to be companion or sporting animals. I found Susanna Forrest's book, *The Age of the Horse*, with its sweep through the history of humanity's relationship with the horse, and, as a geographer, I became intrigued by the question of how horses have shaped places and how our relationship with horses has left its imprint on our geography. Taking Britain as a locus for a case study, what would an historical geography of Britain look like when considered from the perspective of the relationship between humans and horses? Mindful of the recent growth of scholarship in the relatively new field of 'animal geographies' (or sometimes now called 'more-than-human' geography), I looked for literature on the geography of the horse but found very little. So, I set about learning about horses for myself through writing this book. The book has not therefore arisen out of any previous equine expertise. I had no special affections for, nor experience of, the animal although my interest has been stimulated as I pursued my research. The book is instead born out of the inquisitiveness of a geographer who stumbled upon the question of the geography of the equine world and the equine contribution to our more-than-human geography. I hope the book provides a stimulus for others to take on the task of better understanding the equine influence upon our contemporary and historical geography.

Acknowledgements

This is the second book I wrote during a research sabbatical at the University of East Anglia (UEA) after a lengthy period spent on the University's Executive Team embroiled in the challenges of higher education management. I had planned to write a book on food and climate change, and *Net Zero, Food and Farming* had a relatively smooth gestation and birth and so I found myself in February 2022 with some time to spare. It was the first research leave of my academic career, and I faced the unusual situation of having no academic or administrative commitments and a completely free hand to pursue whatever I wanted for a few months. What more could an academic wish for? This book is the result of my rather opportunistic equestrian excursion.

I would like to thank those who made writing the book possible. At UEA, I am grateful to David Richardson and Frances Bowen for providing me with a sabbatical, which meant I could conduct the research for the book. I am also grateful to Mark Searcey, Kevin Hiscock and Ian Renfrew for helping arrange for my move to UEA's School of Environmental Sciences, known within the University by its shorthand moniker of 'ENV'. It was an unexpected but very positive turn of events to find myself joining the School and to be able to engage with its academic staff and students from across the social and natural sciences. ENV has provided an ideal academic community and intellectual home for me to finish the book.

I have benefitted from conversations with many people which informed my thinking. For their time and help, I would like to thank Tom Almeroth-Williams, Sue Bradley, Christopher Brown, Sam Chubbock, Georgina Crossman, Jez Fredenburgh, Bethan Gulliver, Matthew Herbert, Philip Howell, Damien Maye, Roly Owers, Olga Petri, Jeremy Phillipson, Chris Ritson, Jan Rogers, Polly Russell, Mark Wentein and Abi Williams. One of the first people I spoke to about the book was my former colleague, Henry Buller, who was characteristically enthusiastic and supportive, and it is a sad loss that he died before I could send him the finished product and we could talk more. Special thanks are due to Susanna Forrest, whose book, *The Age of the Horse*, was an inspiration and who pointed me to valuable source material and provided some helpful editorial guidance. I am also grateful to the British Equestrian Trade Association and the Royal Agricultural Society of England who gave permission for me to use their data. I would also like to thank Faye Leerink and Prachi Priyanka at Routledge for their help and support. Most of all, I am extremely grateful to those who generously read my draft material. All errors in the text are my responsibility alone, but my thanks go to Tom Almeroth-Williams, Sue Bradley, Sam Chubbock, Gillian Damerell, Joe Darrell, Andrew Donaldson, Steve Dorling, Tamzin Furtado, Bethan Gulliver, Allyson May, Damian Maye, Venetia Morrison, Chris Philo, Emma Roe, Chris Ritson, Lee-Ann Sutherland and Polly Ward.

Researching for the book has introduced me to some wonderful scholarly studies in anthropology, social and economic history and historical geography, and I would like to pay tribute to the following works, which I thoroughly recommend to readers. Tom Almeroth-Williams's *City of Beasts* (2019) was an important source of material on eighteenth-century London that I drew heavily upon in Chapter 4, along with Mark Brayshay's landmark study of *Land, Travel and Communications in Tudor and Stuart England* (2014). Mike Huggins' *Horse Racing and British Society in the Long Eighteenth Century* (2018) is a fine history of the development of horseracing as a popular national sport. Rebecca Cassidy's two books on horsey places and cultures, *The Sport of Kings: Kinship, Class and Thoroughbred Breeding in Newmarket* (2002) and *Horse People: Thoroughbred Culture in Lexington and Newmarket* (2007), both bring the horseracing world alive. Finally, no scholar of the history of the horse in Britain can do without the pioneering work of Peter Edwards, including *The Horse Trade of Tudor and Stuart England* (1988) and *Horse and Man in Early Modern England* (2007).

Abbreviations

BCE	Before Common Era
BETA	British Equestrian Trade Association
BHA	British Horseracing Authority
BHC	British Horse Council
BHIC	British Horse Industry Confederation
BHS	British Horse Society
CE	Common Era
COFICHEV	Conseil et Observatoire Suisse de la Filière du Chevals (Swiss Council and Observatory for the Horse Industry)
Defra	Department for Environment, Food and Rural Affairs
EAFPN	Equine Assisted and Facilitated Practitioners' Network
EBTA	Equine Behaviour and Training Association
HBCA	Human Behaviour Change for Animals
FEI	Fédération Équestre Internationale (International Federation for Equestrian Sports)
FMD	Food and Mouth Disease
MAFF	Ministry of Agriculture, Fisheries and Food
NEF	National Equine Forum
NFU	National Farmers' Union
PTSD	Post-Traumatic Stress Disorder
RSPCA	Royal Society for the Prevention of Cruelty to Animals
UEA	University of East Anglia

1 Horses, Humanity and Scholarship

Introduction

Horses have played a greater role in shaping our world than any other animal. Their influence on human history spans the building of great cities, the clashing of civilisations and the exercise of imperial power. Horses evolved from forest-dwelling creatures the size of small dogs around 55 million years ago. Their descendants spread around the world's land masses and were hunted by humans for food. Eventually domesticated between 5,000 and 6,000 years ago, the horse became a farmed food source and then a draught animal pulling wagons and chariots. At some point, humans worked out how to ride on its back. This not only enhanced people's ability to hunt other animals for food but also radically transformed their mobility. It changed our perspective on the world and our place in it.

Horses became crucial beasts of war and conquest. They were an integral part of Roman imperial power and the Norman invasion of Britain in 1066 and were still prominent features of warfare 850 years later in the early twentieth century. They gradually replaced the ox as the agricultural draught animal of choice. They were exchanged as gifts in international diplomacy. Their use in mills helped power the preliminary rumblings of the industrial revolution. They carried people and hauled freight. They were our principal source of power and their role in enhancing human mobility and interconnectedness is second to none in the animal world. Their contribution to territorial communication and postal services in Britain, as elsewhere across Europe, helped establish a sense of a national population in touch with one another. As clusters of inns, stables, farriers and other businesses built up around coaching stops, so horses and the support infrastructure around them helped shape the evolution of national settlement structures along main strategic routeways and the growth of market towns.

The use of the machine gun during the First World War and the development of electric-powered trams rapidly killed off the use of horses for war and urban transport in the early twentieth century. Their decline in agriculture and

DOI: 10.4324/9781003454359-1

some areas of haulage was a more protracted process over the decades from the 1920s to 1950s, leaving Britain with the horse as primarily a creature of sport and leisure. Three million riders and six million racegoers help support a British equine economy of around £8 billion turnover a year.[1] In the country, horses represent the third most significant land use after food production and forestry. In the city, horses remain prominent figures in displays of royal pageantry. They pull hearses at funerals and still sometimes serve as instruments of crowd control, should a protest march turn nasty.

It is only relatively recently in the humanities and the social sciences that scholarly attention has begun to focus on the place of non-human animals in the evolution of culture, economy and society. The impulse to take animals more seriously in our studies of humanity is an extension of the poststructuralist concern about marginalised others. This means much of the work in animal studies has been about representation and discourse; that is how animals are portrayed or understood. More materially, the ecological crisis and the rise of animal welfare concerns have stimulated consideration of how humans treat animals and with what implications for our health, our sense of ethics and the future of the planet. To date, there has not been so much work in animal studies on how human–animal relations may materially shape places, nor has there been much work by scholars in geography that considers horses. This book is therefore an effort to fill a gap and raise a set of questions about how animals might shape geography. It is a study of the evolution of human–animal relations over time, focusing on the horse in Britain, and the implications for the development of places. It is a study of people, power and place, but from an equine perspective. The book takes Britain as a case study to develop an historical geography of how, with the help of horses, Britain has evolved. It examines historical and contemporary equine geographies of economy and society to understand the current place and prospects of human–horse relations.

Before domestication, horses were wild animals. Following this 'pre-domestication' era, Susanna Forrest writes of *'The Age of the Horse'*, that period of human history in which the horse played a vital and shaping role.[2] Her book is a celebration of the ways human culture and economy have been nourished and formed by the relationship with the horse. From the pre-domestication era, through the rise of the horse and the age of the horse, we reached what we might call 'peak horse' in the early twentieth century in Europe and North America. Ulrich Raulff has written of the end of the age of the horse. In *Farewell to the Horse*, he focuses on what he calls "the final century of our relationship".[3] While we may be a century past the point of 'peak-horse', we are left with a set of uses and places for horses, and a set of equestrian legacies that still leave their mark. Horses are the ultimate marginalised 'other' in our social and economic history. They played a crucial role in the development of so much of human civilisation but have not been sufficiently considered in historical accounts and academic research. The first use of horses to humans was as meat. They were eaten by hunter gatherers and domesticated to eat. They were also milked.[4] Horsemeat has long been used to feed other animals, as pet food, but human consumption of horse meat is met with revulsion in some places including Britain and North America. After meat came other uses, as a source of transport and power. As horse-drawn technology developed through the medieval and early modern period, so human mobility was extended, trade across land expanded and horses thus contributed to not only a strengthened sense of interconnectedness between communities but also an enhanced capacity for conflict and conquest. Through the medieval and early modern period, horses were a vital resource in the comparative power of armies. They played their part in forging nations by force and developing militarism through sport and culture. After the Norman conquest, hunting on horseback developed as a popular pastime among the ruling elite in Britain and a strong link developed between military horsemanship and equestrian pastimes.

Given these important roles in military power, transport, commerce and industry, the production and trade in horses developed to become an industry in itself. Horse breeding became a preoccupation of royalty and the aristocracy, but also a livelihood for farmers and smallholders across rural areas. Accumulating fine horses was a symbol and a measure of wealth and status among the wealthy. Horse fairs and markets became an important engine of trade and economic development in small towns and across rural regions. The gradual development of more regional and national trade and movement in horses led to the growth of favoured horse-trading centres, and so a distinctive geography of the horse economy developed. Wealthy users of horses became more refined and discerning in their demands and more specialist breeds of horses developed, to hunt the fox, haul the plough, race each other or pull little carts down the mine or around the town.

Horses were revered in art and came to be a prominent subject of painting, sculpture and literature. They feature extensively in statues celebrating military might and bravery in Europe's capital cities and civic squares. Outside the British Parliament stands Boudicca of the Iceni, arms aloft on her chariot behind two rearing horses. The twentieth century western movie, a prominent genre from the 1930s to the 1960s, was a mainstay of world cinema and popular culture with its 'Cowboys and Indians' upon their horses. Together, these different features of equestrian life help map out the uses and places of horses in our historical geography and in our popular imagination. Three questions arise. First, what imprints have humanity's relationship with the horse left in the contemporary age? In other words, what is the legacy of the age of the horse? Second, is what of the relationship today? What is the nature of contemporary equine geographies of economy and society, and what are their prospects? Third, is how have research and scholarship gone about understanding the evolving place and roles of animals, and particularly horses, in our geography?

Animal Geographies: Horses and the Social Sciences

The last thirty years have seen marked growth in research in the social sciences and humanities on the relationships between humans and non-human animals. Sometimes called 'human–animal studies', or just 'animal studies', this literature spans history and cultural studies, literary studies, sociology, anthropology and geography. Figure 1.1 provides a simple illustration of the growth publications in 'animal studies' since the late 1980s according to the Web of Science database of academic publications. Numbers gradually rose during the 1970s and 1980s, and then rose dramatically after 1989. Whole journals, such as *Society & Animals*, are dedicated to the subject and it is now commonplace for animal studies to feature in the agendas of the main disciplinary conferences across the humanities and the social sciences.

What explains this growth? Developments in social theory have been an important stimulus. Two broad trends can be identified. The first is the rise of poststructuralist, feminist and postcolonial theoretical perspectives which highlight power relations and the marginalisation or 'othering' of repressed groups, both through socio-economic practices but also through ways of thinking, framing and representing. These ways of producing new knowledge can reveal how prevailing norms, ideologies and power relations are perpetuated or reinforced in ways that include and exclude, privilege and marginalise. This major body of theory has paved the way for whole swathes of social science and humanities scholarship to scrutinise social and cultural relations from a poststructuralist and post-colonialist perspective. The mission is to render power relations and processes of marginalisation and repression visible and so potentially amenable to emancipatory and liberationist action. It has spawned new interdisciplinary communities of scholars around African-American studies, women's studies,

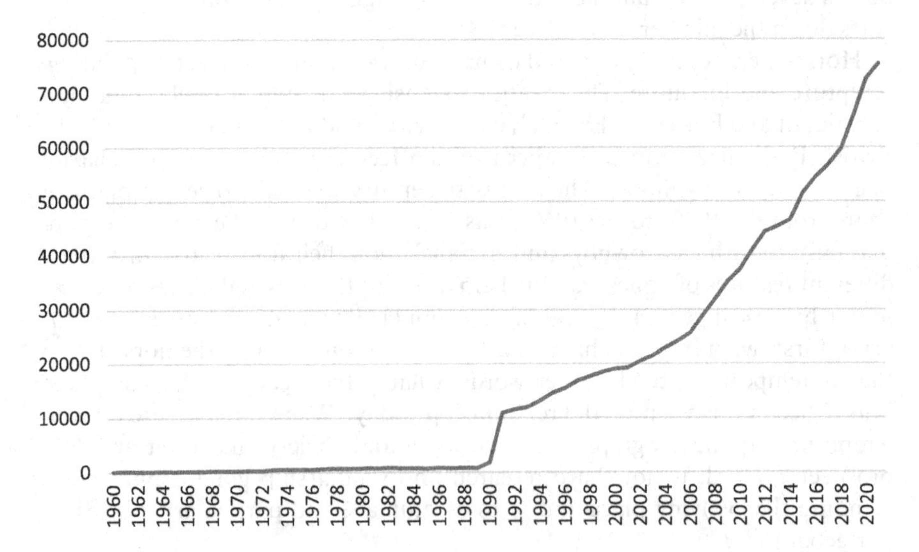

Figure 1.1 Annual Number of Academic Publications in Animal Studies, 1960–2021.

Note: Data derived from the Web of Science Core Collection using the term "animal studies".

disability studies, queer studies and so on. The rise of animal studies can be seen as an extension of poststructuralist social theory, casting animals as another category of marginalised 'other' deserving of critical attention.

A second and related trend is the increasing interest in the 'social' and 'natural', the ways in which these categories are made and maintained, and the implications for social theory and scholarly practice of unsettling and problematising them.[5] This trend is largely stimulated by the ecological crisis and the rise of environmentalism, but also through social science studies of science and technology. Its impact has not been so extensive across the social sciences and humanities. It is less of an influence upon animal studies rooted in political science, for example, but more prevalent within geography, a discipline in which both social and natural scientists find themselves institutionally bundled together. Deconstructing the 'social' and the 'natural' raises questions of what it means to be human, non-human or 'more-than-human'.[6] Humanity has been manipulating plants and animals for millennia, but developments in the biosciences and especially genetics are increasing the scale and pace at which plants and animals can be altered and at the same time opening up new terrains of contestation around what might be right or wrong in humanity's relationships with the non-human world.

These prevailing trends in social theory heavily influence the intellectual work that is done to study human relations with non-human animals. Historians argue that the various foci within history often reflect the prevailing interests of the day. "Historians' sense of what was important in the past tends to mirror their sense of what is important in the present".[7] The rise in animal studies can also be thought of as symptomatic of a contemporary impulse, a product of our times. There is heightened concern about a global crisis in biodiversity as habitats are damaged or destroyed altogether and species become extinct. The expanding global population and the growing demand for meat is fuelling a profound transformation in the agri-food system with existential implications for our socio-economic systems through greenhouse gas emissions. Perhaps the growth in interest in animal studies might also reflect a maturity in the humanities and social sciences, and a confidence to move beyond the human foci of disciplinary conventions and traditions and to address topics previously the more exclusive preserve of the animal scientists in veterinary sciences, biological sciences, ecology and so on.

Among the social sciences, human geography may look hamstrung by its name when it comes to engaging with animal studies, but boundaries have never been an obstacle for this most imperial of disciplines, more a source of fascination. There was once a tradition of studying the distribution of animal species by physical geographers, the natural scientists within the discipline, which was strongly associated with zoology. Its focus on what types of species are found where, on migration patterns and correlations and causations in data, fitted with conventional, geometric and abstracted notions of space. It was part of a technicist approach to statistically analysing patterns and relationships in the animal world that had already withered to become a niche area within geography by the 1950s and had largely faded away by the 1970s.[8] Since then, physical geographers have not been much interested in animals. Their preoccupations have largely been around geomorphology, hydrology, climate and earth systems science. Biogeographers, tessellating with ecology and population biology, have at times been interested in the roles of animal species in ecological systems. Paleoecologists also trace the distribution of microorganisms in lake sediments to understand long term environmental change, though the question of whether the diatom is an animal need not detain us here. In short, the recent growth of animal geographies since the 1990s has not been stimulated by interactions *within* the discipline between human and physical geographers. Rather, it has been through human geographers' encounters with wider debates across the social sciences and humanities around social theory, cultural studies and environmental ethics.[9] Geographers have picked up what is happening elsewhere and generated a new sub-discipline of 'animal geographies'. Figure 1.2 traces the rise in publications (in 'animal geography' or 'animal geographies') since the 1980s using the same approach as for Figure 1.1. The trajectory is similar to that for 'animal studies' but follows around a decade later. Animal geographers were following in the path of other disciplines.

A landmark moment in the birth of the 'new animal geography' was the publication in 1995 of a themed issue of *Environment and Planning D: Society and Space*, edited by Jennifer Wolch and Jacque Emel, under the banner of 'bringing the animals back in'. The rationale was that human geography, and social theory more widely, was hitherto "resolutely anthropocentric".[10] The introductory paper asserted "animals are signifiers, denied lives of their own. Animals are the ultimate Other".[11] The argument was about fairness, the need to build a more inclusive social theory. Capitalism had converted animals into

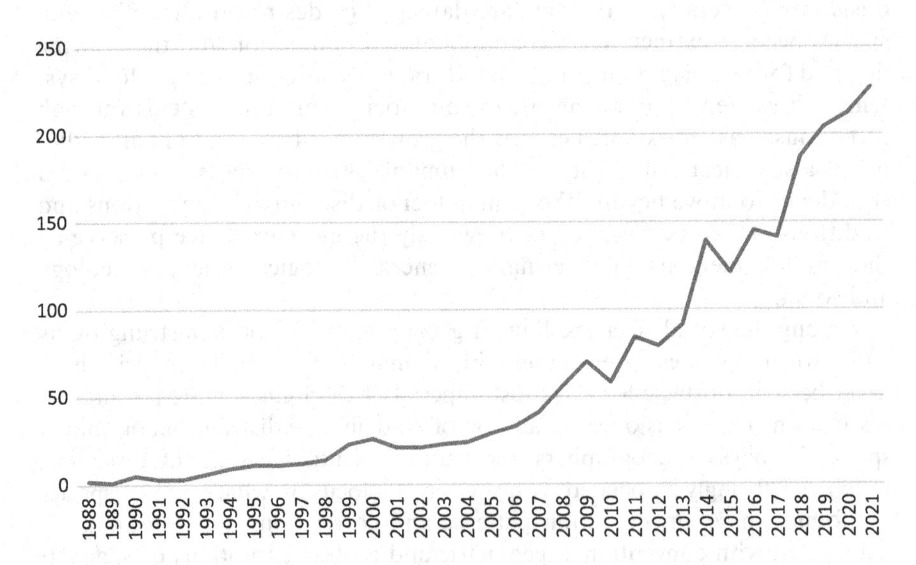

Figure 1.2 Annual Number of Academic Publications in Animal Geography, 1988–2021.

Note: Data derived from the Web of Science Core Collection using the term "animal geography".

commodities. Patriarchy and anthropomorphism had, hand in hand, served to marginalise humans' kinship with non-humans and urban-centric modes of living and thinking were tending to purify space of animals. Animals were being either actively written out and marginalised, or thoughtlessly overlooked. Yet they should be an essential part of our social and geographical theories about how the world works as it does, and how it came to be this way. The editors argued "animals are central to environmental sustainability, economic and social order, personal relations and individual identity, and conceptions of justice and morality".[12]

The response was rapid, and a proliferation of edited collections and agenda-setting manifestos followed.[13] Chris Philo and Jennifer Wolch reviewed geography's engagement with animals over its history and argued that the revival of interest, a 'new animal turn', was an important opportunity to examine

> the complex nexus of spatial relations between people and animals. The goal is to tease out the myriad economic, political, social, and cultural pressures shaping these relations with reference to both particular groupings of people and particular species of animals.[14]

The degree of influence of social theory, cultural studies and critical race and postcolonial theory has stimulated fascinating work in animal geographies including in animals' roles in the development of culture and questions of animal subjectivity and agency. Identity and representation are prominent foci, including "the ways in which ideas and representations of animals shape personal and collective identity".[15] A recent review by Alice Hovorka and colleagues celebrates more than 25 years of animal geographies.[16] The field has "challenged humanist ontologies and epistemologies and advanced multispecies theoretical and empirical understandings in geography".[17] They argue for a stronger emphasis on power, life-worlds and praxis and issue a rallying call: "This is the moment for more explicitly pointing to the interconnectivity of animals-humans-environments, as well as highlighting the experiences of marginalised human and nonhuman groups".[18]

Animal geographers have worked on all sorts of animals — wildlife, farm animals and pets. Research has focused on zoos and farms, on animal welfare issues and on representations of animals in popular and media discourses. A rich literature has developed on cats, dogs, elephants, mice, cougars, dolphins and lots of other creatures. A scan through the index of a recent edited book collection lists references to cows, cockatoos, bees and bonnet-macaques.[19] The (bio)diversity is spectacular, and commendably so, but from the perspective of this study, it begs the question what of the horse? Perhaps it is a privileging of the exotic over the mundane. Perhaps it is a contemporary association of the horse with more affluent middle-class people and privilege. Whatever the reason, the animal geographers' 1990s charge of the "unexamined absence" of animals from geographical scholarship can just as easily be turned on animal geographies almost 30 years later when it comes to horses.[20]

A search for geographical work on horses reveals few results. One study from the 1980s by R.W. Tomlinson at Queens University Belfast described the distribution and development of flat-racecourses in Britain from their origins in the sixteenth century.[21] A further study by Phil McManus at the University of Sydney and colleagues is of the global horseracing industry, looking at trends in the industry, its role in the development of some key localities and ethical issues emerging around the sport.[22] Another study from 2013, which fits more within the tradition of the new animal geographies, is by Cheryl Nosworthy based on her doctoral research at the University of Reading and considers issues of affect and emotion in the interactions between horse and rider in the context of disability and recovery from a horse-riding accident.[23] It is a fascinating and intimate analysis of human–horse relations and makes a significant contribution to disability studies. It does not, and was not intended to, shed much light on the broader questions of the historical evolution of human–horse relations nor the contemporary political economy of the horse. Catherine Nash at Queen Mary's University of London has looked at the relationship between place and animal breeds[24] and has examined horse breeds and kinship in Iceland.[25] In 2021, Nigel Thrift produced *Killer Cities* examining how cities cause mass animal death and suffering and hasten the destruction of the planet and he spends a few pages on horses.[26] He argues that city infrastructures are mainly designed without animals in mind, a point that could be debated especially from the historical perspective where many European and North American cities were heavily moulded around the horse in the eighteenth and nineteenth centuries. For material questions that ask how many horses, from where, doing what, with whom and with what implications for people, power and places, animal geographies is not the place to start. There is an equine paradox at the heart of animal geographies in that the species which probably forms the most significant and globally extensive human–animal relationship in the history of humanity has not been the focus of much work.

Beyond geography, the horse does feature within animal studies in other parts of the humanities and social sciences. Compared to geography, there is quite a literature on the history of the horse, although even there, writers are critical of its limited scope. Histories can be found of the evolution of horse breeds and uses, the evolution of knowledge about horses,[27] and of the role of horses in warfare,[28] policing,[29] and mining.[30] Global history "has thus far developed very little horse sense", writes Ulrich Raullf.[31] While globalisation and increasing geographical interconnectedness have been dominant themes of social science over recent decades, political and economic historians have paid very little attention to what Raulff calls "the central vector of every historical land power, the horse".[32] However, rich new studies have emerged in history and cultural studies of the changing relationship between humanity and horses. Donna Landry has written of the ways in which particular breeds of horses helped transform British equine culture and sense of self and empire.[33] Social historians such as Tom Almeroth-Williams have traced the roles of horses in the development of cities.[34] Popular anthropological writings have dealt with how culture and economy have been influenced by horsepower,[35] and historians have traced in a broad arc the rise and fall of the age of the horse.[36] Yet these are not geographical studies. Recent work in history, anthropology and cultural studies has enriched our understanding of the importance of the horse in culture, economy and power, but there has been little focus on the nexus of horses, power and place. In horse studies there has been a lack of thinking about geography, while in animal geographies there has been a lack of thinking about horses.

What distinctive contribution can geography make to animal studies? In addressing this question, Chris Philo and Jennifer Wolch set out two broad traditions in human geography. The first concerns questions of space, distribution and location while the second concerns place, region and landscape.[37] Traditionally, geographers studied distribution and patterns of phenomena across different locations. This form of Cartesian 'spatial science' involved the search for spatial laws to explain patterns and distribution, to explain why things were where they were. The mobility of people and things adds an additional dimension and the rise of computer power means there is plenty to study in the patterns of migration, commuting, supply chains and so on. Since the 1970s, 'spatial relations' have been subject to vigorous conceptual debate among human geographers and others, and it has become acknowledged that spatial relationships are themselves socially constructed. Distributional patterns, and the relationships between people and things across space, are a function of social, economic and political processes and practices that shape our world and shape places within it. The social construction of space opens questions of the relationships between powerful places and elsewhere, what might be called the cores and peripheries, how places come to be more similar or more different in their characteristics and how people and things might be included or excluded as spatial relationships evolve. A contribution that geographers can make to animal studies is therefore in understanding the spatialised nature of relationships between humans and non-human animals.

The second tradition is in the study of place, that is specific localities which might be a village, a town or city, a region, nation or continent. The rise of spatial science and its quest for generalisable laws in the early post-war decades rendered the study of particular places or regions unfashionable for a period, but over recent decades there has been renewed interest in the importance of place and sense of place, to economic development, social wellbeing and environmental sustainability. Places can also be thought of as socially constructed. Their current form is a function of both past history and contemporary geography. Doreen Massey developed a geological metaphor for understanding changing places, pointing to historical waves of social, economic and technological changes, each leaving a layer of imprint or sedimentation on localities.[38] At the same time, and throughout history, places will be being shaped by interactions with elsewhere.

Territorial struggles mean that in many parts of the world places 'change hands' between different ruling forces. Flows of trade, people and ideas, by sea or over land, shape particular places at particular times. Geographers study how places change and develop, and for animal geographers there is the opportunity to examine how animals, and the relationships between humans and animals, might influence the development of places, landscapes or regions. To date, however, this has not been a prominent preoccupation of work in animal geographies. The main impact of animal geographies has been on questions of social theory, ontology, epistemology and ethics.[39] Because animal geographies developed after the poststructuralist turn in human geography in the 1980s and early 1990s, the principal questions have come to centre around the extent to which animals might be marginalised, or 'spoken for' in social science analyses. While this is all well and good, it has meant that there has not been as much focus on the material implications of animal geographies and the ways that animals might play roles in the social and material construction of places. So how might a more-than-human geography of the horse be approached?

The Book's Approach

The approach to this study of the historical and contemporary geographies of human–horse relations is rooted in the recent evolution of contemporary human geography, mindful of the influence of poststructuralist theory including its role in helping bring animals back into our studies of nature society-relations. It takes a dynamic and relational view of space and place, which sees our contemporary geography as a heterogeneous make up of spatial formations that include, mix up and reformulate the 'natural' and 'social', the human and non-human. Although informed by poststructuralist thought, the study seeks to understand more than just how horses and their worlds are talked about and thought about. It addresses the material as well as the representational, the historical as well as the contemporary, the animal as well as the human, but all through the prism of geography, places and power.

Poststructuralism has given rise to a relational view of space and place in which they emerge out of dynamic processes of change. "Space is generated by interactions and interrelations".[40] That is to say, "space can no longer be seen as simply a 'container' of heterogeneous processes; rather space is now thought to be something that is (only provisionally) stabilised out of such turbulent processes, that is, it is made by heterogeneous relations".[41] Space and place are made, are made relationally and are thus conceptualised as "territories of becoming that produce new potentials".[42] Such a perspective allows new spatial imaginaries that bring together relations, representations and the "fleshy materialities of the socio-spatial world".[43] A relational approach helps address how space and place evolve over time through the changing relations between people, other places and resources. David Harvey conceptualises place formation as the establishment of what he calls permanences. He writes:

> The process of place formation is a process of carving out 'permanences' from the flow of processes creating spaces. But the 'permanences' – no matter how solid they may seem – are not eternal: they are always subject to time as 'perpetual perishing'. They are contingent on the processes that create, sustain and dissolve them.[44]

As yet, work in animal geographies has had relatively little to say about the role of non-human animals in place-making. This study contributes through a geographical analysis of the evolution of the relationship between humans and horses and the implications for a place, Britain, in the contemporary period. It is a study of the evolution of social and spatial processes over time but centred on the specific role of the horse as a non-human actor in the geography of a nation. It places the horse at the centre of an analysis not only of the internal development of Britain, but also of British relationships with other places, including the relationships of trade, conflict and domination by the world's first global imperial power.

Britain is a good site for a case study of the relationship between horses and humanity for several reasons. As an island nation, the need for the transport of horses in and out of the country by sea has rendered these flows more visible, documentable and measurable than they might otherwise have been. That Britain was a pioneer of the agricultural revolution and the first country to go through the rapid industrialisation and urbanisation of the industrial revolution in the eighteenth and nineteenth century also provides a useful and distinctive prism through which to assess the importance of the horse-human relationship and its geographical implications. A key technology that eventually helped humanity transition away from dependence upon the horse was the railway, which was first developed in Britain, although its use spread very rapidly worldwide from the mid-nineteenth century. The decline of empire, deindustrialisation and the particular and enduring attractions of rural living, even in an age of globalisation and post-industrial urban cosmopolitanism, also make the British case a distinctive site for exploration of the changing place and meaning of the horse in the twenty-first century.

A central concept is that of power, which has become an increasing concern within animal geographies over recent years.[45] Sandra Swart writes that "the history of horses is the history of desire — and the desire for power in particular".[46] Horses provide power in a simple sense of brute strength to haul and carry. In addition, the human–horse relationship is bound up with power relations between people and places. Individual people, collective social groups and classes, and particular places have derived power, material resources and influence through their relationships with horses. Horses shape the changing fortunes of people and places and, as actors, have influence and effects themselves, even though there might be uncertainties about the degree of intentionality that might be ascribed to their actions. Horses make a difference and have made a difference to how places develop.

The book is in four parts. The first part explores the evolution of the relationship between horses and humanity from domestication through early modern times to the industrial revolution and the peak in horse numbers in Britain in the early years of the twentieth century. Over the history of civilisation, humans have used the horse more than any other animal and the relationship was a vital one for several millennia. Until the nineteenth century, horses were the principal means of transportation over land and so underpinned the development of spatial relations between people and places. The use of horses and the infrastructure of the horse economy and transport system prior to the industrial revolution have left its mark on the landscape and built environment as well as human culture. Chapter Two traces the relationship between humanity and the horse up until the industrial revolution. It examines the principal uses of the horse and the role horses played in social hierarchies and power relations in human societies. It considers the different uses to which horses were put, including for power and transport, for racing, for work and for war. It provides an account of the social and institutional infrastructure for producing horses, including horse breeding and dealing. The chapter takes us to the point of the industrial revolution, a pivotal era in humanity's relationship with the horse, after which came the modern period and ultimately the shift in the horse in Britain from a creature of work and power to one more largely of leisure and play. Chapter 3 then examines the relationship between humanity and horses during the process of the industrial revolution and in the century that followed.

The second part of the book takes a more geographically differentiated perspective and considers the evolving relationships between humans and horses in different types of places, and especially in urban and rural areas, and in the process of nation-building and place-making. Chapter 4 first examines the role of the horse in the radical enhancement of mobility and the flow of people and goods around Britain, especially from the sixteenth to the eighteenth centuries. The chapter examines the ways that geographies of industrialisation, urbanisation and deepening interconnectedness all interrelated, but around the central relationship between the human and the horse. It goes on to examine in more detail the roles of the horse in the worlds of work and of play in the eighteenth-century city before considering the implications for the shape of cities. It concludes by reflecting on the de-horsification of the city in the twentieth century. Chapter 5 then provides a parallel account of how the relationship between horses and people has evolved with particular emphasis on work and leisure in the countryside. The chapter describes the role of horses in agriculture and the significance of hunting with horses and hounds in the social structure of rural Britain before examining the influence of horses in the management and maintenance of the rural landscape. It concludes with an assessment of the changing relationship between humans and horses since the advance of mechanisation of agriculture from the mid-twentieth century.

The third part of the book considers the contemporary equine economy and society in Britain. Chapter 6 looks at the recent evolution of the 'horse economy'. It maps the contemporary nature of the horse world focusing on horses in sport and leisure. It delineates the size and trajectories of different components of the horse economy and traces the spatial relationships of the horse economy within the UK and internationally. Chapter 7 considers the contemporary social geography of the horse. It examines who are the people who own horses, ride horses and make up 'horse society'. It explores the contemporary institutions of the horse and the politics of managing the equine affairs of the nation.

The final part of the book considers the evolving nature of knowledge about horses and the prospects for the human–horse relationship. Chapter 8 considers the different realms of expertise about horses and how they have developed over recent decades. It examines the nature of the equine veterinary profession before exploring recent developments and debates about horse breeding and animal welfare. New scientific knowledge is redefining our understanding of the early domestication of the horse and of equine genetics and breeding. Neuroscience is also transforming understanding of the changing nature of human–horse relationships and the role of the horse in human wellbeing and as an instrument of therapy.

Chapter 9 presents the conclusions of the study. It draws out a broader analysis of the changing nature of the human–horse relationship in the twenty-first century. It considers the shift in practice and meaning, from horses as principally agents of transport, agriculture, industry and war to creatures of sport and leisure. It assesses the role of the horse in the formation of places within the UK and in the formation of the nation as a whole. It also reflects on the implications of this historical and contemporary equine geography for animal geographies and animal studies more widely and makes the case for the study of animals in general and horses, in particular, in relational place-making.

Notes

1 British Horse Council (2019) *The British Horse Sector – Why it Matters for the 2019 General Election*. High Wycombe: British Horse Council.
2 Forrest, S. (2016) *The Age of the Horse: An Equine Journey through Human History*. London: Atlantic.
3 Raulff, U. (2018) *Farewell to the Horse: The Final Century of Our Relationship*. London: Penguin.
4 Wilkin, S. *et al.* (2021) Dairying enabled Early Bronze Age Yamnaya steppe expansions, *Nature* 598, 629–633.
5 Latour, B. (1993) *We Have Never Been Modern*. Hemel Hempstead: Harvester Wheatsheaf.
6 Whatmore, S. (2002) *Hybrid Geographies: Natures, Cultures, Spaces*. London: Sage.
7 Ritvo, H. (2002) History and animal studies, *Society & Animals* 10, 403–06.
8 Philo, C. and Wolch, J. (1998) Through the geographical looking glass: Space, place and society-animal relations. *Society & Animals* 6, 103–18.
9 Emel, J. *et al.* (2002) Animal geographies, *Society & Animals* 10, 407–12.
10 Wolch, J. and Emel, J. (1995) Bringing the animals back in. *Environment and Planning D: Society and Space* 13, 632–6.
11 Wolch and Emel (1995), p.632.
12 Wolch and Emel (1995), p.632.
13 See, for example, Philo, C. and Wolch, J. (1998); Philo, C. and Wilbert, C. (2000) *Animal Spaces, Beastly Places: New Geographies of Human-Animal Relations*. London: Routledge; Buller, H. (2013) Animal geographies I, *Progress in Human Geography* 38, 308–18; Buller, H. (2014) Animal geographies II: methods, *Progress in Human*

Geography 39, 374–84; Buller, H. (2015) Animal geographies III: ethics, *Progress in Human Geography* 40, 422–30.
14 Philo and Wolch (1998), p.110.
15 Emel *et al.* (2002), p.408.
16 Hovorka, A. *et al.* (eds.) (2021a) *A Research Agenda for Animal Geographies*. Cheltenham: Edwin Elgar.
17 Hovorka, A. *et al.* (2021b) Introduction, pp.1–20 in Hovorka *et al.* (2021a), p.2.
18 Hovorka *et al.* (2021b), p.3.
19 Hovorka *et al.* (2021a).
20 Wolch and Emel (1995), p.632.
21 Tomlinson, R. (1986) A geography of flat-racing in Great Britain, *Geography* 71, 228–39.
22 McManus, P. *et al.* (2013) *The Global Horseracing Industry: Social, Economic, Environmental and Ethical Perspectives*. London: Routledge.
23 Nosworthy, C. (2013) *A Geography of Horse-Riding: The Spacing of Affect, Emotion and (Dis)ability Identity through Human-Horse Encounters*. Newcastle upon Tyne: Cambridge Scholars Publishing.
24 Nash, C. (2020) Breed wealth: Origins, encounter value and the international love of breed, *Transactions of the Institute of British Geographers* 45, 849–61.
25 Nash, C. (2020) Kinship of different kinds: Horses and people in Iceland, *Humanimalia*, 12(1), 118–144.
26 Thrift, N. (2021) *Killer Cities*. London: Sage.
27 Budiansky, S. (1997) *The Nature of Horses: Their Evolution, Intelligence and Behaviour*. London: Weidenfeld & Nicholson.
28 Ellis, J. (2004) *Cavalry: The History of Mounted Warfare*. Barnsley: Pen and Sword Books.
29 Campbell, J. (1967) *Police Horses*. Newton Abbott: David & Charles.
30 Thompson, C. (2008) *Harnessed: Colliery Horses in Wales*. Cardiff: National Museum of Wales.
31 Raulff (2018), p.346.
32 Raulff (2018), p.346.
33 Landry, D. (2009) *Noble Brutes: How Eastern Horses Transformed English Culture*. Baltimore: Johns Hopkins Press.
34 McShane, C. and Tarr, J. (2007) *The Horse in the City: Living Machines in the Nineteenth Century*. Baltimore: Johns Hopkins Press; Almeroth-Williams, T. (2019) *City of Beasts: How Animals Shaped Georgian London*. Manchester: Manchester University Press. See also Sleeman, P. (2023) *Bricks, Stones & Straw: Working Horses in Liverpool*. Stroud: Amberley Publishing.
35 Forrest (2016).
36 Raulff (2018).
37 Philo and Wolch (1998).
38 Massey, D. (1984) *Spatial Divisions of Labour: Social Structures and the Geography of Production*. London: Macmillan.
39 See the following progress reports for overviews of the preoccupations of animal geographies over recent years. Hovorka, A. (2017) Animal geographies I: Globalizing and decolonizing. *Progress in Human Geography* 41, 382–94; Hovorka, A. (2018) Animal geographies II: Hybridizing. *Progress in Human Geography* 42, 453–62; Hovorka, A. (2019) Animal geographies III: Species relations of power. *Progress in Human Geography* 43, 749–57; Gibbs, L (2020) Animal geographies I: Hearing the cry and extending beyond. *Progress in Human Geography* 44, 769–77; Gibbs, L (2021) Animal geographies II: Killing and caring in times of crisis. *Progress in Human Geography* 45, 371–81.
40 Murdoch, J. (2006) *Post-Structuralist Geography: A Guide to Relational Space*. London: Sage, pp.19–20.
41 Murdoch (2006), p.4.

42 Thrift, N. (2004) Summoning life, pp.81–103 in P. Cloke, P. Crang and M. Goodwin (eds.) *Envisioning Human Geographies*. London: Arnold, p.88, quoted in Murdoch (2006), p.17.
43 Murdoch (2006), p.17.
44 Harvey, D. (1996) *Justice, Nature and the Geography of Difference*. Oxford: Blackwells, p.261.
45 Hovorka (2019).
46 Swart, S. (2010) *Riding High: Horses, Humans and History in South Africa*. Johannesburg: University of Witwatersrand, p.viii.

2 Horses in Pre-Industrial History

Introduction

J. Edward Chamberlain has written that "horses have had more influence on the rise and fall of civilisations than any other factor".[1] The domestication of the horse therefore represents a pivotal moment in human development with profound consequences across large parts of the world. Horses originally provided meat and milk, have been a source of status and authority and played a prominent part in human culture and arts, but it is primarily for their sheer power that humans have harnessed and made use of horses. The origins of the domesticated horse lie in the wild horses of the steppes of central Asia, covering southern Russia, Kazakhstan and Mongolia. Archaeological evidence suggests domesticated horses were present in Europe, west of the Ukraine by 2000 BCE (before common era), and large-scale horse breeding was likely to have begun in the first millennium BCE.[2] By the late Bronze Age, domesticated horses are relatively abundant in the archaeological record.

Humanity's relationship with the horse has shaped the human geography of much of the world. Until the industrial revolution, horses were a principal means of transportation over land and thus vital in the development of spatial relations between people and places. Horses underpinned human mobility and were crucial in the developing geography of inland interconnectedness. Travel by horse resulted in a network of roads and bridleways still discernible today. By the end of the seventeenth century, the horse-drawn coach had become an established form of territorial transport across Europe, bringing the development of roadways and coaching inns.[3] Horses played an essential role in the transmission of news and the ways in which people came to understand what was happening elsewhere. Horses were essential to the strength and effectiveness of armies and thus an indispensable part of state power.

The use of horses and the infrastructure of the horse economy and transport system prior to the industrial revolution have left its mark on the landscape and built environment as well as on culture. This chapter explains the relationship between humanity and the horse up until the industrial revolution of the late eighteenth and early nineteenth century. It explains the origins and domestication of the horse and its effects on human civilisations. It considers

DOI: 10.4324/9781003454359-2

the power relations surrounding the horse, and the development of the horse economy and horse breeding during the early modern period.

The Evolution and Domestication of the Horse

More is known about the evolutionary origins of the horse than any other animal. It evolved from a small, dog-sized forest-dwelling creature with four toes that lived over 55 million years ago, *Sifrhippus*. The evolution of horses largely took place in North America and there is evidence of migration of equids between the North American and Eurasian land masses over land bridges. Different lineages evolved in different places and sometimes several could co-exist. There was a significant wave of dispersal from North America to Eurasia at the time of the *Merychippus* genus, which were the first equids to enter Africa around 10 million years ago. The smallest horses, known as 'dawn horses', became extinct on the Eurasian continent around 23 million years ago but continued to thrive in North America, grew in size and evolved hooves from the central toe. They were the size of small ponies and existed at a time when the first human-like primates, the *hominin*, appear in the fossil record.

By 4 million years ago, the horse was becoming distinguishable from the donkey and zebra and around two and a half million years ago, early horses spread from North America to Latin America and across Eurasia dwelling on grasslands. They began to encounter humans, emerging from Africa, initially as a new predator. By 500,000 years ago, horses lived across the world's American and Eurasian landmasses and prehistoric humans were hunting them with spears. The main animal types were the cabelines (the main wild horses, including the Tarpan and what is known as Przewalski's horse) and the non-cabelines (the ass and zebra lineages). Although still a relatively diverse set of species, the wild horses of the time had some common characteristics and were smaller and heavier-headed than the horse we know today.[4]

Across Europe, there is archaeological evidence that Neanderthals scavenged horse carcases, although by 55,000 years ago there is much more extensive evidence of active hunting. Horse bones are found among prehistoric human settlements, with concentrations in southern France and northern Spain. By 30,000 years ago, humans were demonstrating a keen interest in the horse, with images of horses dominating palaeolithic rock art. Painted images of horses on cave walls give us some sense of what the horse, *Equus caballus*, looked like at that time and the extent of the relationship with humans. A study of over 4,700 instances of animal depictions in cave art across Europe from between 30,000 and 12,000 years ago showed that horses feature in three-quarters of sites and account for 30 per cent of all animal depictions.[5] Prehistoric humans seem to have been fascinated by the horse. At the Solutré site in southern France there is evidence of major horse kills from the palaeolithic period where a limestone ridge was used to corral horses. Kills took place regularly over a 20,000-year period. After the land-bridge between the Americas and Eurasia was submerged under the sea, the horse gradually disappeared from the Americas. This extinction, which took place around 10,000 years ago, was likely to have been the result of either climatic changes or over-exploitation by humans. In Europe, horses became more heavily hunted and with the growth of woodland they migrated south and west, especially to the grassland steppes of central Asia and Siberia.

Our understanding of the domestication of horses has shifted over recent years with significant new research findings published in 2021.[6] This remains a highly fluid area of scientific knowledge, with a lively scientific debate about the precise point in time and space where domestication may have first occurred. Ancient horse types that have at times been thought of as the source of domestic horses include the Tarpan (*Equus caballus gmelini*), a local variety of wild horse on the steppes, which became extinct in the late nineteenth century. In Mongolia, a tawny coloured wild horse, known after the Russian geographer and explorer, Nikolai Przewalski, as Przewalski's horse (*Equus caballus przewalski*), and also called the Takhi, was 'discovered' by Przewalski during an expedition to Central Asia in 1878. The Takhi has also been thought to be a possible ancestor of domesticated horses and a small herd of these wild horses continues to survive, bred and reintroduced from captive animals from zoos.[7]

Over recent decades, there has been increasing interest in the hypothesis that the domestication of the horse took place first among the Botai culture in northern Kazakhstan around 3,500 BCE. The Botai had been hunter-gathers but had settled into a semi-sedentary settlement structure that seems incompatible with hunting wild herds. Archaeological finds suggest they were heavily dependent on horses and horsemeat. Over 90 per cent of the bones found at Botai sites were from horses. Horse dung was used to insulate the thatched roofs of their buildings and horse jawbones were used as tools to make leather thongs from horse hides. Fatty acid deposits found on Botai cooking pots very likely derive from mare's milk, suggesting horses must have been domesticated. This evidence is reinforced by the fact that dental analysis suggests that the Botai horses were bridled and bitted.[8] It is not clear whether the Botai rode domesticated horses to hunt wild horses, but they certainly kept horses.

Recent evidence has emerged to suggest that today's domesticated horses originated in the western Eurasian steppes and particularly the lower Volga-Don region of what is now southern Russia bordering on Ukraine. This new research is presented as resolving longstanding debates about the origins and spread of the domestic horse. Over 270 ancient horse genomes were examined and showed that domestic horses from the Volga-Don region replaced almost all other local populations over a period of just a few centuries up to between 1,500 BCE and 1,000 BCE after they were domesticated and rapidly bred to meet growing demand.[9] The expansion of these horses was associated with the spread of the Indo-Iranian language and the use of chariots.[10] Horses were selectively bred to develop back strength and docility in this period. The Botai may have been the first to domesticate the horse, but the Botai horses are not the origin of today's domesticated horses.

The nomadic tribes of the steppes, which stretch more than 6,400km from Hungary in the west to China in the east, were able to develop as horseback warriors and roam further and further afield and began to harass the more sedentary communities that had settled along agriculturally fertile fluvial areas. The horse enabled the development of long-distance over-land trade routes which in turn enabled trade and exchange of goods, but also brought cultural mixing and the diffusion of new ideas, religions and technologies, including technologies associated with horse-riding and haulage. Horse-based warfare wrought destruction, but in its wake horsepower "greatly extended the scale and complexity of civilisation".[11] Cultural anthropologist, Pita Kelenka, has traced the evolution of human–horse relations across 6,000 years and points to several major migrations, enabled by the horse and of progressively broader scale, which shaped the development of the modern world.

The horse culture that developed among the Indo-Eurasian agriculturalists of the steppes over the fourth and third millennia BCE enabled expansion to the west into Europe and eastwards across the Eurasian steppes. The horse, and the development of wheeled vehicles to accompany it, meant the agro-pastoralists became much more mobile and were able to range across larger territories extending for thousands of kilometres in contrast to the sedentary and circumscribed states that were developing in Asia and North Africa at this time. The heightened mobility of these increasingly horse-based cultures and the development of war chariots meant military domination of the urban centres of sedentary civilisations in the Near East, North Africa, Iran, India and China. In the second millennium BCE, the Hittite peoples of Anatolia, now Turkey, used horses to drive out vanquished populations. By the first millennium BCE, the recurved composite bow-and-arrow facilitated the development of the warhorse to complement the battle chariot and so spawned military cavalry. Ancient Greek historian, Polybius, who studied the warfare of the third century BCE, concluded that it was more effective to fight with twice as much cavalry as the enemy and only half as many infantry than to enter battle with equal numbers of each.[12]

The Medes, an ancient people based in what is now Iran, selectively bred larger horses to carry armed warriors, an innovation which enabled more extensive "equestrian empires" to grow and absorb the older nuclear states.[13] The Achaemenid Empire from 550 BCE to 330 BCE stretched from the Balkans and eastern Europe in the west to the Indus Valley in the east, while the Qin in the east occupied much of what is now China. In these two empires of the first millennium BCE, extensive road networks were built to facilitate horse-based transport and unified languages and written systems were imposed. By the end of the first millennium BCE, the war chariot and mounted cavalry were ubiquitous as the power behind the state across Eurasia and North Africa. Kelenka writes:

> the introduction of the domesticated horse into Europe … ushered in a period of rapid social change and technological progress; it also resulted in the new formation of cultural groupings from which the Indo-European languages – Greek, Italic, Celtic, Germanic, Albanian, Baltic, and Slavic – would evolve.[14]

The horse became revered and celebrated, including in the amphitheatres of Rome.[15] It underpinned trade along the Silk Road between Occident and Orient and helped diffuse the ideas of Buddhism, Judaism and Christianity. Warring equestrian invasions continued from the steppes, and in the sixth century CE, Mongolian Avars introduced the metal stirrup to Byzantium which enabled horse-backed warriors to bear lances. Soon after, Byzantine dominance of the Mediterranean began to be challenged by equestrian forces from the Middle East and North Africa. By the end of the first millennium CE, the religious strife driven by divisions in the new Islamic world was exacerbated by Seljuk incursions from what is now Turkey and the (horse-born) crusades from Western Europe to the Middle East.

Invasions from the steppes to the west or from the deserts to the south had only ever led to partial conquests across Eurasia. However, in the thirteenth century, a much stronger equestrian force emerged from Mongolia whose conquests ranged from the Pacific to the Baltic, and eventually led to a deeper exchange of technical know-how and luxury goods. Printing, gunpowder and forged steel flowed west from China to Europe. The Mongols were eventually deposed in the fifteenth century by the rise of the Ottoman Empire, powered by Turkish horses, which invaded Europe as far west as Vienna. Muslim control of Middle Eastern trade routes helped prompt the western European powers to turn to the seas to further develop trade. Notably, at the end of the fifteenth century, Columbus introduced a batch of war horses to the Americas. They were followed by hundreds more which eventually enabled the conquering of more than 25 million people across the Americas where the horse had been extinct for 9,000 years. Pita Kelenka argues that the horse served as the technological bridge from the middle to the modern age, and the rapid transport and communication that the domestication of the horse facilitated ranks alongside agriculture and metallurgy as one of the most important innovations in human history.

In Britain, it is likely that horse-breeding began on a large scale in the first millennium BCE. In prehistoric times, the horse seems to have been endowed with sacred characteristics which may have been because of its speed or its resistance to taming when compared with more placid ruminants. In the early Iron Age, slender-limbed ponies were used to pull wheeled carts in Britain, and it is likely that their early use in transport was for military purposes. The early Celts used chariots and the horse is the most common animal to feature on Celtic coins.[16] Julius Caesar invaded Britain in 54 BCE reportedly accompanied by five legions and 2,000 cavalry, with a further three legions and 2,000 cavalry left over the Channel in Gaul. Cavalry proved pivotal in overcoming early resistance by the Britons.[17] In 2023, a group of scientists announced archaeological research findings that established for the first time that Viking invaders of Britain in the ninth century were likely to have brought horses with them in their long boats as they crossed the North Sea from Scandinavia. Previously, it had been assumed that the Vikings had pillaged their horses from the local Saxons, but the new scientific evidence, based on analysis of strontium isotopes in cremated Viking remains found in Derbyshire, suggest that the Vikings' horses had been brought from Baltic regions.[18]

Horses were not used extensively in agriculture until centuries later. For a horse to pull a plough horizontally with all its strength required a harness and collar that would not press on the windpipe. This technology was developed in central Asia and not brought to western Europe until around CE 700. Ploughs were first pulled by oxen, and a tenth-century Welsh law even required this to be so. Horse-breeding looks to have taken place at some Roman villas in England which served as studs and where horse remains are abundant in the archaeological record. During Saxon times, horses and whole stud farms were left in wills, showing evidence of an organised horse-breeding industry. Some 1,100 years after the Roman invasion, horses again played a significant role in the next invasion of Britain, this time by the Normans. William the Conqueror invaded in 1066 with an army including 2,000–3,000 horses and overpowered the defending forces at the Battle of Hastings. Horses featured extensively in the battle and almost 200 are depicted in the Bayeaux Tapestry which tells the story of events.[19] By the time of the Domesday survey following the Norman conquest, there was around one riding horse per village. Feral and unbroken horses lived in woodlands, but their numbers declined considerably in the decades after the Norman conquest, most probably because of the extensive forest clearance that took place. As elsewhere in the world, populations of feral ponies persisted in parts of the British countryside, including the New Forest, Dartmoor and Exmoor, the Pennines and Wales. These feral populations are descended from horses that were previously domesticated, rather than truly wild varieties.

Horses, Power and Status

From ancient civilisations through early modern times, horses both embodied and reflected power. With their bodies, they could exert power to haul carts or carry riders, but horse ownership also reflected power and status among individuals, families, tribes and nations. It has been calculated that by the late eleventh century, over 70 per cent of power in English society was provided by animals.[20] By the sixteenth century, although most people did not own a horse, a proportion of the peasant-labouring population did. Horse ownership was most common in the north, where 40 per cent of cottage farmers kept horses, and least common in the eastern counties, where the proportion was 12 per cent. In the Midlands, closer to horse-markets like Stratford upon Avon or the manufacturing areas like Birmingham, some better-off labourers began breeding horses in response to the growing demand for packhorses in these areas. Typically, though, a sixteenth century farm-labourer's horse was a multifunctional beast serving the same diverse sets of purposes as the peasant's donkey in Ireland. Inventories and lawsuits from the early modern period show how the labourer's horse was used in "carrying turf and bracken in packs, panniers and 'nets', slung across its back; in carting hay, corn, and wood; and in conveying the labourer's wife and neighbours to the nearest market town".[21]

Owning horses to ride rather than to work became a point of social differentiation and those better off labourers who began to prosper, often through their work as horse breeders and traders to meet rising demand, were able to derive prestige from their conspicuous ownership of saddled horses. We might say horse-ownership became a middle-class thing during the sixteenth and seventeenth centuries. Probate inventories demonstrate a rise in horse ownership during the seventeenth century.[22] For example, one-fifth of householders in Yetminster in Dorset owned horses in the 1590s, but this proportion had grown to three-fifths by the 1660s. Parish rates records show that the proportion of ratepayers who owned horses rose from three in five in 1636 (60 per cent) to almost three-quarters (73.4 per cent) in 1701, over four in five (81.1 per cent) in 1724 and to six in seven (85.7 per cent) by 1742.[23]

For the upper classes, riding on horseback not only provided utilitarian mobility but also social prestige, and good horsemanship became a key signifier of social status. The quality of horses, in terms of their size, shape and colour, became of interest and value and the numbers of horses owned helped calibrate social status. It would be common for local gentry to keep ten to twelve horses, but nobility would have dozens and their stables would be conspicuously grand. At the top of the social hierarchy, the monarch would own hundreds of horses and Henry VIII, for example, owned over a thousand around the country at the time of his death.[24] The more horses owned, the more had to be spent on their care. Numbers employed might range from one or two stablemen to the Crown estates where staff employed to look after horses ran into hundreds. There was also the question of the land required to grow the feed. In 1562 it was calculated that it cost five shillings a week to feed a horse in service, requiring hay, straw, oats and peas. By 1702, the cost had more than tripled.[25] Among animals, horses enjoyed a special status, and one that was physical, psychological, and cultural as well as material. Peter Edwards and Elspeth Graham write:

> Possession conferred status and, as a result, horses were imbued with an iconic significance. If early modern society could not have functioned effectively without horses, nor could its human population have understood or engaged with the world in many of the ways we have come to associate with the period without their association with horses. [26]

Horse-riding became a leisure pursuit for the upper classes as well as a mark of distinction and refinement and a symbol of aristocratic rule. Public displays of horsemanship took place through the *manège*, an arena for practising and showing off horse-skills which became a popular elite sport and pastime during the reign of Elizabeth I. The relationship with the horse became a metaphor for governance. The methods to train and handle a horse informed wider approaches to weaponry, personal conduct, land management, administration and government, and eventually to gender relations.

Books were written on the art of good horsemanship and riding schools developed around the country to train young gentlemen in the necessary skills.

Interest in the manège grew across Europe, especially in Paris, and an academy was established in London. The relationship between horses and people both reflected and helped constitute the gender relations of the early modern period in Britain, both in terms of material practices and through representation in art and literature. Although women and children did ride, horse-related activities were generally run by men. Parallels were drawn between breaking a colt and training it to obey its rider and 'training' women to learn socially acceptable behaviour. Male supremacy was implicit in Christian thought, with men exercising authority and providing for their families while women were 'reined in' at home bringing up children.[27] There were widespread fears of breakdown of the social order in the late sixteenth and early seventeenth century and ritual admonishment and humiliation of perceived deviants. Disobedient wives or 'hen-pecked' husbands would be publicly shamed by being set upon a horse and ridden through the streets, in the latter case with the man facing backwards or dressed in women's clothes.

Horses played a central role in the exercise of military power, both in terms of the battlefield, supporting military logistics and through the representational power of iconographies of militarism and equestrianism. The battles of the late medieval period had begun to demonstrate how well-drilled and equipped infantry could, where the terrain was favourable, defeat heavily armoured knights on horseback, as longer pikes could outreach lances. During the sixteenth century, the development of firearms such as muskets changed the nature of battle so that the proportion of cavalrymen in armies declined and infantrymen rose. Mounted troops still performed valuable military functions in battle, although these were more about protecting infantry rather than making the traditional frontal assaults. They could help launch surprise attacks, intercept convoys and scout. Following the development of the wheel-lock pistol in the mid-fifteenth century, many cavalry units effectively became mounted pistoleers. In the seventeenth century, the cavalry charge came back into military vogue, usually following an artillery barrage and musket salvo and the development of cannons meant a couple of horses could pull lighter, more mobile guns.

The political economy of horse production and ownership was heavily shaped by the needs of the Crown. Henry VII had legislated in 1495 to ensure that his subjects were obliged to fight for king and country in times of war. Sixteen years later, Henry VIII surveyed his leading subjects to identify how many men they might be able to muster. In particular, the landed gentry were expected to provide horses. In any civil war or insurrection, the ability to raise horses was critical, and any lack of support was damaging. The upper classes gravitated towards the cavalry, whose exclusivity was reinforced by the cost of the horses and their accoutrements. The county militia, under the leadership of a lord lieutenant, the monarch's representative in the English shires, provided an additional source of soldiers, horses and military equipment. Tudor monarchs formalised the militia and developed a system for the supply of matériel and attendance at musters in times of conflict. After 1558, what was required of subjects was set according to a sliding scale of wealth. Those at the bottom end of the scale had to keep some small arms including a long bow, sheaf of arrows and a steel helmet. Those at the top of the scale contributed six demi-lances and ten light horses, together with their mounts and arms, as well as additional arms for thirty to forty further soldiers.[28]

The Crown's requirements of the early modern militia were a source of tension and conflict, with much argument about levels of financial and material assistance, attendance at musters and whether substitutes could be sent. Levels of compliance with obligations varied across shires and the relatively poor condition of the English militia horse became a national concern. As rebellion rose in northern England in 1569, there were efforts to tighten up on abuses and a revised book of rates for supplying horses was produced. Yorkshire supplied eighty-two rated horses, with a further 172 donated. Other shires continued to grumble. The perceived general lack of cooperation and worsening international situation prompted the establishment of a powerful Special Commission to further tighten up the system of supply of horses for warfare and putting down rebellion. The Commissioners imposed quotas on some counties, but difficulties persisted.[29] During the English Civil War of the seventeenth century a new dynamic was introduced. Control of county militias was a key potential source of fighting men and horses, and supplying resources became a touchstone of loyalty to the cause. Despite expectations that enough horses would be raised, both warring factions had to introduce elements of compulsion. Eventually, both sides found sufficient horses for their military purposes. The increased demand spurred breeders to expand production, but it took four to five years for new horses to be ready for battle.[30]

Draught horses and wagons to support military forces were acquired differently from cavalry horses. Henry VIII tended to use his royal privilege of purveyance to appropriate horses either at fairs or parish by parish. However, taking horses from farms or the road had knock on consequences for loss of production. Henry's early military adventures were challenged by the poor quality of English equine resources but they prompted a complete transformation in the quality of native stock in general and in the horses required by the army in particular. By the end of the sixteenth century, English horses were admired and in demand abroad.

Horses and the Early Modern Period

By the early modern period, the horse had come to be a mainstay of everyday life for many people and a prominent feature of farms, towns and villages. People's uses of horses had become more diverse, including not only a range of utilitarian purposes but also for frivolous pleasures. Animals in general, and horses, in particular, were viewed through an almost totally anthropocentric view and their purpose was solely to serve humanity, although this idea began to come under some challenge and questioning.[31] The livelihoods of many thousands of people depended on horses and horse-related industries, and the breeding and feeding of horses became a significant part of the rural economy.[32]

The wealthier classes employed servants to look after their horses, and blacksmiths, farriers and horse leeches earned their living providing equine services. Stables became a significant and sometimes prominent part of larger farms and houses, and their construction and maintenance provided work for bricklayers, carpenters, labourers and so on. The closed fields in which horses grazed had to be managed with hedges and fences maintained. Whole industries developed to produce saddles, harnesses and other essentials, and cartwrights, ploughwrights, wainrights and wheelrights worked to manufacture and repair ploughs, harrows, carts and wagons.

In agriculture, horses shared work with oxen, mules and asses and were to become the dominant form of haulage and traction power. Ploughs had generally been drawn by oxen, or sometimes horses or both. Oxen were more likely to be used on tougher land, but horses began to be seen as more effective on lighter soils. Horses were more expensive to keep, and held less value at the end of their lives compared to oxen, which could be sold for meat.[33] A crucial issue was the availability of fodder rather than the cost of draught animals. Oxen required lush grass and could not graze common fields in the ways that horses could. Mixed farmers could keep oxen, although in the uplands horses would also be required. Common-field farmers in the lowlands were more likely to stick to just horses. The relative merits of oxen and horses were much debated but gradually, over time, the horse began to take over as British agriculture's dominant method of traction.[34]

The growth in horse ownership and use in the sixteenth and seventeenth centuries altered the pattern of flows of people and things around Britain and so shaped the development of socio-spatial relations, the ways people and places interacted with each other. Horses could be hired from a hackneyman, whose numbers grew during the sixteenth century as population growth stimulated domestic demand. Businesses started to hire out horses stipulating that passengers must stay in their particular inn. Where journey speed was important, a chain of horses could be laid out, although such relays of horses were expensive and only available to the wealthiest. In 1603, Sir Robery Carey rode from Charing Cross in London to Holyrood Palace in Edinburgh covering 400 miles in less than sixty hours, including overnight rests.[35]

The development of postal roads was designed to hasten travel by horse. The first route was from London to Dover, followed by a route north to Berwick. Early in his reign, Henry VIII appointed Brian Tuke as the first Master of the Post and tasked him with developing a system of staging posts so that royal dispatches could be swiftly conveyed. A postmaster was stationed at each staging post. The system represented the first example of an integrated national terrestrial transport network and was centred on the efficient exploitation of the horse. By the late 1560s, further routes had been added totalling 420 miles and with some 40 post rooms.[36] By the end of the sixteenth century, postal roads radiated out from London, terminating at Dover, Falmouth, Milford Haven, Holyhead, Carlisle and Berwick, and covered 1,200 miles of road and 79 post rooms. Branch routes later evolved to develop a much more extensive network.[37] By the time the network extended over 1,500 miles, there were scarcely any places in England located more than twenty or thirty miles from a staging post.[38] As geographer Mark Brayshay put it, this "enhancement of England's communications infrastructure by means of the network of royal and general standing posts represented a major victory over distance and spatial isolation".[39] Private passenger traffic was encouraged alongside post on these routes and after 1660 postmasters were obliged to supply a horse to customers within half an hour. Postmasters often served as innkeepers as well and were active in local schemes to maintain a supply of horses to supply this increasingly national network of horse-dependent interconnectedness.

Horses carried out a widening range of tasks over the early modern period across agriculture, transport and haulage and in brewing. Over time, horse breeders began to more carefully consider the specific traits required for different tasks. In transporting goods, the choice was between small pack ponies or larger draught horses. The former could carry only around a third of the load of the latter but were better able to deal with hill country or rough or muddy ground. Packhorses were around a third more expensive than draught horses and by the end of the seventeenth century, carrier services to London were split evenly between the two options. Transport costs helped determine the geographical pattern of industry. Because transporting goods by water was much cheaper than by horse, coal mining developed more intensively in mines located close to navigable rivers such as in Durham and Northumberland. There horses would cart coal to the wharfs with a single horse pulling a two-wheeled cart with almost 9 hundredweight of coal per load. Horses were used to cart other goods to and from navigable rivers, and the river port of Lechlade in Gloucestershire developed during the seventeenth century as the navigable limit of the River Thames where cloth and cheese from the surrounding areas could be loaded and transported by river to London. Goods could be hauled more efficiently by horses if vehicles were either on rails or floating on water, and horses were used to haul barges upstream against the current or wind. It was estimated in 1730 that six horses could do the work of more than sixty men hauling goods by barge. Wagonways were gradually developed first in Nottinghamshire, then the north east and Shropshire, which greatly increased the loads of coal that could be hauled. The first two-mile track in Nottinghamshire enabled at least an extra 180 tons of coal to be moved a week.[40]

As agriculture and industrial production began to become more regionally specialised, so greater volumes of goods and material were transported over longer distances. Twenty cartloads of corn would be taken from Dorset and Wiltshire to Shaftsbury each market day by the early seventeenth century and in the north a single farmer would usually send eight loads of grain at a time to market in Malton in Yorkshire. Similarly malt and cheese would be transported to larger urban markets or to ports for onward shipment to London. Wool and finished cloth were transported long distances by carriers too. Carrier services were provided by a range of businesses. Common carriers serving London were often also farmers who owned land, produced fodder and other crops, and had relatively large numbers of horses. They were professional people of some standing in their communities. One carrier, William Claroe, based in Worcester, in 1680 had twenty-three horses, four wagons, a tumbrel and a cart. Other staff would normally be hired to do the driving. Those operating at shorter distances within a thirty-mile radius tended to be of more modest means and often smaller farmers who would cart loads using their farm horses at quieter times of year. Packmen often tended to have just one or a few horses, although those carrying over longer distances would need more horses. A list of carriers serving London in 1681 contained five who would journey between thirty and fifty-four miles who on average owned five horses and seven who travelled between 105 and 146 miles had nine horses. One packman, Joseph Naylor of Rothwell, had over a hundred horses.[41]

The two-wheeled wain or cart was the traditional vehicle used to haul goods, but from the time of Elizabeth I these began to be replaced by larger four-wheeled wagons which arrived from the continent and whose use gradually spread across Britain. They were initially more prevalent in lowland areas but by the eighteenth century had spread to the north and west. They also began to be used on farms, again spreading from the south east and initially among gentlemen and more prosperous yeomen farmers. They were most commonly used for taking grain to market. In the period 1670 to 1690, slightly under one in ten farmers left wagons in their wills.[42] In the uplands, packhorses were more likely to be preferred to carts and wagons. The adoption of larger wagons stimulated demand for stronger horses. The Suffolk Punch with its thickset body and short legs was well-suited as a workhorse and larger and stronger horses also came from around the Fens and Somerset Levels. Although it took some time, the average size of English draught horses began to grow in the early eighteenth century through breeding.[43] After the Glorious Revolution of 1688 which put William of Orange on the throne, more horses were imported from the Netherlands which contributed to the improvement in England's stock of draught horses, although the professional long-distance carriers were quicker to obtain stronger draught horse than farmers.

Horses were also used to power horse engines, or gins, which could power a milling machine, threshing machine or lifts. At coalmines, gin gangs were used to wind up coal and to drain pits by drawing up water. The numbers of horses involved in mining work could be large. For example, in 1602, Lord Willoughby had sixty-five pump horses to drain his pits at Wollaton in Nottinghamshire.[44] Flooding could be extremely disruptive of coal mining and cause severe local hardship and so horsepower for pumping played an important role in keeping coal mines functioning. Gin horses were also used to draw water from wells for human consumption or to power machines to draw water from rivers.

The use of horses on farms, roads and mines expanded considerably over the two centuries to the 1750s and by this time the horse had virtually wholly taken over from oxen as the beast of burden. The growth in horse use was partly as a result of an expanding population and the resultant growth in agricultural production and industrial output, but also because more land was brought into production and manufacturing was beginning to expand. Production of food and manufactured goods became more specialised and this also generated the need for more material to be moved about. Horse use became more specialised too and this shaped the ways that horses were bred and traded.

Producing Horses

Until the 1980s, there had been little research into Britain's pre-industrial horse trade, despite their important role in social and economic life. Historian Joan Thirsk remarked:

> When we consider the age-long dependence of man on the horse coupled with the dramatic expansion of economic activity in the early modern period, it is remarkable how little interest historians have shown in the way that horses were made available to meet more insistent and fastidious demands.[45]

The horse trade became more regulated during Tudor times and transactions had to be recorded in toll books. These were similar in purpose to a modern vehicle log-book and contained information on prices, numbers and types of horse sold, their size and function. Analysis of these toll books by historian Peter Edwards provides a detailed empirical picture of the trade supplying horses in the early modern period. Much horse breeding and rearing was initially local and small-scale, with animals reared and sold among neighbours and at local markets and fairs. Most horses were multi-purpose, helping with tasks on farms and on roads. However, as more people began using horses and for more specialised purposes, this stimulated the emergence of more specialised breeding and rearing areas, and the gradual development of the horse market on a national scale, including the bringing in of foreign blood to improve horse lines. Although the dominant imperative in importing horses came from the upper classes and their obligations to provide cavalry horses for the national army, the demand for better quality horses extended more widely than for military purposes alone. Over time, imported horses and selective breeding gradually improved the national stock bringing benefits across the economy and society.

In the early part of the Tudor and Stuart period, horses were traded at markets and fairs and trading was highly localised. Even in the late seventeenth century, toll books show that over three-quarters of sellers and more than two-thirds of buyers at Brewood in Staffordshire lived within ten miles of the town.[46] Fairs tended to be founded on the fringes of horse-rearing areas and operated throughout the year, but with December and January as quieter months.[47] Horses did begin to be traded over longer distances, but this was generally by passing through several hands over time rather than through single trades. There was a general drift of horses to eastern and south-eastern England from the breeding areas of the north, west and south west, although underlying this were more complex local patterns. Areas around the Fens were also prominent horse breeding sites, but horses from these places can be traced westwards and then eventually to the south east. The upper classes employed agents to find horses that met their wider range of more specialist needs and so tended to purchase from further afield. Much of the longer distance trade passed through the Midlands where a greater concentration of fairs developed, including nationally renowned centres at Northampton and Penkridge.[48]

Horse breeding was carried out on farms, usually where there was sufficient pasture land for horses to be fed relatively cheaply. Some farmers bought young stock and concentrated on rearing and training horses to work and selling them on as mature draught horses. In the early sixteenth century, a large proportion of the English horse population were ponies of fourteen hands or smaller, bred on farms across the country and often roaming in a semi-wild state until they were broken in. Heavier horses were more likely to be bred and reared in the areas with improved grasslands and the largest of all came from the fenland areas. The Fens were home to a distinctive wetland environment that underpinned a particular farming system. Subject to significant drainage and land reclamation schemes during the seventeenth and eighteenth centuries, they eventually became a prime arable farming area.[49] It was the availability of grazing land and supply of hay that suited them to breeding and rearing horses.

Henry VIII became concerned at the state of English horse stock and sought to legislate to ensure there were greater numbers of larger and stronger horses. His father, Henry VII, had banned the export of serviceable horses in 1495 and an Act of 1540 banned the population in 25 specified counties from keeping stallions over the age of two and below fifteen hands on any common or waste ground. All mares, fillies, foals and geldings thought to be too small to either produce foals of sufficient stature or provide useful labour were to be destroyed. There were also concerns about the movement of English horses into Scotland and the punishment for taking horses into Scotland was first increased to a fine of £40 in 1547, then a year's imprisonment in 1549 and was made a capital offense by 1580.[50] Over the sixteenth and seventeenth centuries, the stock of larger horses did increase, but numbers from smaller breeds thrived too.

Some of the most popular horses came from Scotland and Wales such as the Galloways from south-west Scotland or Montgomeryshire horses. Defoe described Galloways as the best breed of "strong low horses in Britain, if not in Europe ... These horses are remarkable for being good pacers, strong, easy goers, hardly, gentle, well broke, and above all, that they never tire".[51] Herds of semi-wild horses would graze on Welsh common pastures and be rounded up at the age of three and brought to market. These horses from Wales and the Welsh borders were drawn into areas of industrial development in the north and mingled with those from Scotland. Fairs at Carlisle in the north and Bridgenorth, Ludlow and Shrewsbury in the west supplied dealers serving the north and the west Midlands.

Breeding and rearing the larger dray and saddle horses was well suited to the areas on the fringes of the Fens in Lincolnshire and East Anglia. Larger colts were also bred on the Somerset Levels. Most renowned of the larger draught horses was the Suffolk Punch, which originated from south-east Suffolk, although similar strains could be found in Norfolk. The relatively low position of the shoulders on a Suffolk Punch meant the horse could throw more of their weight into the collar and so provide greater power. In the north, the Cleveland Bay had been bred to become a larger and stronger draught horse. Evidence from toll books suggests that larger horses began to be traded over longer distances and a national trade began to develop. Somerset dealers took larger horses from the Levels to Winchester where they were sold on to rearers from Wiltshire, Hampshire and Sussex. They were also moved northwards to Staffordshire and Leicestershire for rearing in the east Midlands, which by the 1720s had become England's foremost rearing area. Defoe wrote that these horses:

> are the largest in England, being generally the great black coach horses and dray horses, of which so great a number are continually brought up to London, that one would think so little a spot as this of Leicestershire could not be able to supply them.[52]

A national network developed, extending from Somerset to Yorkshire and East Anglia, and centred in the east Midlands, as the breeding ground for producing England's biggest and strongest horses. Northampton came to be the national centre of the horse trade and, with Rothwell, between Northampton and Leicester, became a major source for metropolitan dealers serving the London market. Between 1684 and 1720, a third of horses sold at Rothwell were to Londoners.[53]

Henry VIII's concerns about the quality of the national stock of horses persisted, not helped by his own military adventures which depleted stocks. During the war against France in the 1540s it was found that four Dutch horses could do the same work as seven English ones, and some thought this was the result of young English horses being worked too early. Importing foreign stock was seen as one means of improving the qualities of native breeds. German and especially Flemish horses could improve English stocks of heavy cavalry and draught horses. Danish draught horses were admired and could also pull a plough, cart and coaches. Coursers from Naples became popular as heavy cavalry mounts, especially when cross-bred with native mares. Eastern and North African stock — Arabs, Turks and Barbs — began to be imported and were well-suited to the new cavalry manoeuvres that required speed and agility. These horses were also raced and their cross-breeding with English mares was the source of the modern racing thoroughbred.

Exchanging fine horses as gifts became part of Henry's international diplomacy. He began to build up royal racing stables, although when purchasing foreign horses he most commonly bought draught horses from the Low Countries. In 1525 he bought 300 horses in Holland and in 1544 he purchased 200 mares from Flanders.[54] Royal imports were an important influence on the breeding of horses, but the gentry were also expected to set a good example in their counties by keeping a good stud. This could be challenging because of the costs involved. Breeders like Sir Roger Pratt of Ryston Hall in south west Norfolk were careful in managing the economics of high quality horse-breeding. In the 1670s, he would only keep the best colts for the saddle and the cart and sell on all others at the age of three.[55] The growing pool of serviceable animals produced by upper class breeders became a valuable national resource for use in war and served the national interest. For those of more modest means, increasing demand ensured there was still a profit to be made from horse-breeding, although they were less likely to take the initiative to improve stock through selective breeding and the use of imports. Improved stock from the upper classes did gradually work their way through to the horses owned by the wider population. Some people could buy horses directly from local estates, and the gentry's surplus horses were sold into local fairs.

By the end of the eighteenth century, a national system of producing and trading horses was in place. There had also been national governmental efforts to seek to strategically develop the stock of horses as a national resource, driven primarily by military concerns. These sought to actively import stock that, through cross-breeding, would positively develop beneficial traits, economically and militarily. They also sought to actively discourage breeding from smaller and less useful stock. The steady advance of heavy black draught horses can be traced through the records of horse fairs of the day. At a fair near Northampton in June 1627 there were more bays and greys than black, but by the end of the seventeenth century at nearby Rothwell over two-thirds of the animals sold were black. By the end of the seventeenth century, English all-purpose horses were increasingly regarded, both at home and abroad, as among the best to be found.[56]

Conclusions

The evolution of the horse as a species is a process that extends over a period of tens of millions of years, with North America an important site but with waves of migration of wild horses around the world's landmasses in response to climate and ecological change.[57] Humans and their forerunners were hunting and eating horses 500,000 years ago, and by 50,000–60,000 years ago there is evidence of horses being extensively revered by humans, and this fascination is reflected in the record of cave and rock art still visible today. Wild horses were domesticated in the steppes of the central Eurasian landmass around 5,000 to 6,000 years ago, initially for food. When setting foot on the Moon in 1969, Neil Armstrong spoke of a small step for a man but a giant leap for mankind. Thousands of years earlier, a human being somewhere took the small but brave step of mounting and riding a horse and so set in train a relationship that was to transform human mobility, warfare and give momentum to the eventual development of modern (more-than-) human societies.[58] We are still learning through carbon dating and genetic research of the early movement of domesticated horses, but the development of horse-riding and breeding and chariot-based warring and raiding are closely bound up with the migration of ancient peoples of central Eurasia across the Eurasian landmass and especially into Europe. The movement of these riders shaped the spread of Indo-European languages and the modern world. The horse is a central and pivotal figure in the development of the world's (more-than-) human geography of the past 5,000 years.[59] It helped transform conceptions of space, place and interconnectedness.

Accounts of the development of the early modern world, the development of agriculture and early industrial production, the rise of towns and cities and the evolution of settlement structures have not tended to centre on the horse. Ulrich Raulff writes:

> Most philosophical, political and sociological theories of space are devoid of that genuine human or non-human vector that crosses the space and opens it up .... Only a few authors have recognised that space is not present until there is motion within it: it is from this motion that spatial relationships develop. It is the battle that creates the battlefield, just as the sailor creates the headland, the rider the bridleway, the climber the ascent, the pedestrians the pedestrian zone.[60]

The movement of people, facilitated by horses, from the Bronze Age onwards brought trade and technological change. Owning horses became an important measure of power and standing, and the horse stock gradually came to be conceptualised as a key military asset and vital to the fighting strength of nations. For example, Henry VIII saw the quality of the breeding stock of horses in England as a key national strategic priority and endeavoured to ensure it improved over time, with some success. Total horse numbers in England rose from around 500,000 in 1300 to over 1.2 million by 1800.[61] As well as a crucial feature of the power struggles between nations, horses played a central role in the development of internal trade, communications and a sense of national identity and belonging. They were the internet of the early modern period. And as their roles and value to people became more diverse, so they began to be actively bred for different valued attributes. A world of horse breeding and trade developed to underpin what was increasingly becoming a horse economy and a horse society.

Notes

1. Chamberlain, J.E. (2007) *Horse: How the Horse Has Changed Civilisations*. Oxford: Signal, p.69.
2. Ryder, M. (1981) Livestock, pp.301–410 in S. Piggott (ed.) *The Agrarian History of England and Wales Volume I. I Prehistory*. Cambridge: Cambridge University Press.
3. Urry, J. (2007) *Mobilities*. Cambridge: Polity Press, p.91.
4. A recent summary of equine palaeontology can be found at: Thomas, B. (2021) The truth of horse evolution – Part 1, Youtube See also Franzen, J. (2010) *The Rise of the Horse: 55 Million Years of Evolution*. Baltimore: Johns Hopkins University Press.
5. Thomas, B. (2021).
6. See the following reviews of the evidence: Warmouth, V. *et al.* (2012) Reconstructing the origin and spread of horse domestication in the Eurasian steppe, *Proceedings of*

the National Academy of Sciences 109, 8202—06; Taylor, W. and Barrón-Ortiz, C. (2021) Rethinking the evidence for early horse domestication at Botai, *Nature* (Scientific reports), 11, 7740; Klecel, W. and Martyniuk, E. (2021) From the Eurasian steppes to the Roman circuses: A review of early development of horse breeding and management, *Animals* 11, 1859. Major new scientific evidence emerged in Librado, P. *et al.* (2021) The origins and spread of domestic horses from the Western Eurasian steppes, *Nature* 598, 634–40.

7 See Forrest (2016) for an account of the history and recent plight of wild horses.

8 Outram, A. *et al.* (2009) The earliest horse harnessing and milking, *Science* 323 (5919), 1332–35; Forrest, S. (2016) *The Age of the Horse: An Equine Journey through Human History*. London: Atlantic, p.16.

9 Librado *et al.* (2021).

10 Anthony, D. (2007) *The Horse, The Wheel and Language: How Bronze-Age Riders from the Eurasian Steppes Shaped the Modern World*. Princeton: Princeton University Press.

11 Kelekna, P. (2009) *The Horse in Human History*. New York: Cambridge University Press, p.2.

12 quoted in Sidnell, P. (2006) *Warhorse: Cavalry in Ancient Warfare*. London: Continuum, p.ix.

13 Kelekna (2009), p.3.

14 Kelekna (2009), p.166.

15 Hyland, A. (1990) *Equus: The Horse in the Roman World*. London: Batsford.

16 Ryder (1981), p.398.

17 Sidnell (2006), p.32 & p.34.

18 Löffelmann, *et al.* (2023) Sr analysis from only known Scandinavian cremation cemetery in Britain illuminate early Viking journey with horse and dog across the North Sea, *PLoS ONE* 18(2): e0280589

19 Lawson, M. (2007) *The Battle of Hastings 1066*. Stroud: The History Press; Bachrach, B. (1985) On the origins of William the Conqueror's horse transports, *Technology and Culture* 26, 505–31.

20 Langdon, J. (1986) *Horses, Oxen and Technological Innovation: The Use of Draught Animals in English Farming 1066–1500*. Cambridge: Cambridge University Press, p.20.

21 Everitt, A. (1967) Farm labourers, pp.396–465 in J. Thirsk (ed.) *The Agrarian History of England and Wales Volume IV 1500–1640*. Cambridge: Cambridge University Press, p.416.

22 It should be noted that these inventories were only drawn up for people with sufficient personal possessions, so proportions are not of the population as a whole.

23 Edwards, P. (1988) *The Horse Trade of Tudor and Stuart England*. Cambridge: Cambridge University Press, p.9, quoted in Edwards, P. (2007) *Horse and Man in Early Modern England*. London: Hambledon Continuum, p.3; see also Thirsk, J. (1984) *The Rural Economy of England: Collected Essays*. London: Hambledon Press.

24 Edwards (2007), p.3.

25 quoted in Edwards (2007), p.4.

26 Edwards, P. and Graham, E. (2011) Introduction, pp.1–33 in P. Edwards, *et al.* (eds.) *The Horse As Cultural Icon: The Real and Symbolic Horse in the Early Modern Period*. Leiden: Brill, p.4.

27 Mendelson, S. and Crawford, P. (1998) *Women in Early Modern England, 1550–1720*. Oxford: Clarendon.

28 Edwards (2007), p.154.

29 Edwards (2007), p.161.

30 Edwards (2007), pp.164–5.

31 Thomas, K. (1983) *Man and the Natural World: Changing Attitudes in England, 1500–1800*. London: Allen Lane.

32 Thirsk (1984).

33 Thirsk, J. (1967) Farming techniques, pp.161–255 in J. Thirsk (ed.) *The Agrarian History of England and Wales Volume IV 1500–1640*. Cambridge: Cambridge University Press, p.164.
34 Langdon (1986).
35 Mortimer, I. (2023) *Medieval Horizons: Why the Middle Ages Matter*. London: Bodley Head.
36 Austen, B. (1978) *English Provincial Posts, 1633–1840*. London: Phillimore. The system was not a new idea and the Romans had developed an earlier version.
37 Edwards, P. (2007), pp.78–9.see also Brayshay, M. (2014) *Land Travel and Communications in Tudor and Stuart England: Achieving a Joined-Up Realm*. Liverpool: Liverpool University Press.
38 Brayshay, M. (1991) Royal post-horse routes in England and Wales: the evolution of the network in the later-sixteenth and early-seventeenth century, *Journal of Historical Geography* 17, 373–89.
39 Brayshay (1991), p.374.
40 Edwards (2007), p.188.
41 Edwards (2007), p.192.
42 Edwards (2007), p.196.
43 Edwards (2007), p.198.
44 Edwards (2007), p.204.
45 Thirsk (1984), p.375.
46 Edwards (1988), p.21.
47 Chartres, J. (1985) The marketing of agricultural produce, pp.406–502; J. Thirsk (ed.) *The Agrarian History of England and Wales Volume V-II 1640–1750*. Cambridge: Cambridge University Press, pp.436–7.
48 Edwards (1988), p.22.
49 Overton, M. (1996) *Agricultural Revolution in England: The Transformation of the Agrarian Economy 1500–1850*. Cambridge University Press; Meeres, F. (2019) *The Story of the Fens*. Stroud: The History Press.
50 Edwards (1988), p.26 & p.47.
51 quoted in Edwards (1988), p.27.
52 quoted in Edwards (1988), p.35.
53 Edwards (1988), p.36.
54 Edwards (1988), p.41.
55 Edwards (1988), p.44.
56 Edwards (1988), pp.50–51.
57 Franzen (2010).
58 The point is made by historian Ann Hyland, quoted in Raulff, U. (2018) *Farewell to the Horse: The Final Century of Our Relationship*. London: Penguin, p.328.
59 Anthony (2007).
60 Raulff (2018), p.340.
61 Wrigley, E. (2006) The transition to an advanced organic economy: half a millennium of English agriculture, *Economic History Review* 59, p.460.

3 Horses, the Industrial Revolution and Empire

Introduction

The industrial revolution set in train a long transition that eventually brought the end of the age of the horse. It was a technological revolution in manufacturing processes that had its origins in Britain but transformed industrial production across western Europe and North America and eventually around the world. Revolutionary change was at its most intense during the period 1780 to 1820. For Eric Hobsbawm, to say that the industrial revolution 'broke out' is to say that "sometime in the 1780s, and for the first time in human history, the shackles were taken off the productive power of human societies".[1] From the 1760s, statistical measures of industrial production took a sudden sharp turn upwards. In what was "probably the most important event in world history", it was as if the economy became airborne.[2] Although industrialisation was taking place in most European countries, revolutionary change was both most advanced and most catalytic in Britain. It was Britain's industrial revolution, coupled with the development of the British Empire, which was to be so significant for the pattern of economic organisation around the world. Technological change was rapid, and economic growth was profound, but the industrial revolution's implications for the horse were drawn out over a 150-year period. Over this period, horse numbers in Britain almost trebled before eventually falling back to their pre-nineteenth century levels.

The conditions in eighteenth century Britain were propitious for an economic revolution as the social foundations had already been laid. A capitalistic agricultural system had been established, with agriculture largely producing for the market. Politics and government were geared to facilitating private profit. The cotton industry was "the capitalist industry par excellence"[3] and had developed in the hinterlands of the major port cities of Bristol, Glasgow and Liverpool. Cloth produced in Britain grew from 20 million yards in 1796 to 347 million in 1830. Manchester, a major industrial city in the cotton trade, saw its population grow six-fold between 1773 and 1824.[4] Britain also had coal, a major source of industrial power in mining and hauling in which horses played such an important part. By 1800, Britain was producing approximately 10 million tons of coal a year or about 90 per cent of world output, and more than

DOI: 10.4324/9781003454359-3

ten times the amount of its nearest competitor, France.[5] Coal-mining helped stimulate the development of the railway which would, alongside the development of the steam engine, bring about a radical change to the relationship between humans and the horse, although this would take many years to unfold.

This chapter examines the roles of horses during the industrial revolution and the development of the British Empire. It explains the gradual introduction of alternatives to horsepower in industrial manufacturing, hauling goods and transporting people. Although conventional historical narratives convey a relatively simple replacement of horsepower by steam power in factories, the railways and the internal combustion engine in transporting goods and people, these processes were not straight-forward, and horses endured in their work with humans long after the use of the new technologies became widespread.

The Industrial Revolution and the Age of Steam Power

As we saw in Chapter 2, horses had become a crucial source of power in transporting people and goods through the late medieval and early modern period and had become an important source of power in some industrial processes, especially in mining, brewing and processing some raw materials. In the late eighteenth century, British industrial production took off, principally triggered by a set of technological innovations. However, the industrial revolution is not a simple story of technology-driven transformation. The social, cultural and material environment in Britain at the time of the revolution is important in understanding why the revolution was experienced in a particular time and place. Horsepower had played an important role in shaping that context and so, although not a prominent feature of accounts of the industrial revolution in Britain, the horse does have its place in the explanation of this world-changing set of events.

The most widely accepted accounts of the industrial revolution place great store by the invention and spread of a set of key technologies and processes. The steam engine, cotton spinning machinery and the use of coal and coke to manufacture iron brought transformational changes in industrial processes that helped radically improve the productivity of the British workforce. A set of other transformational changes in Britain's human geography of followed, including rapid urbanisation, capital accumulation, increased agricultural productivity and income growth. These are associated with the industrial revolution but tended to be consequences of technological change. Although in themselves they had significant implications for economy, society and spatial relationships, they were essentially first order effects of the revolution rather than causes. The expansion of the early modern economy, facilitated by the growth in use of horses, had produced a set of circumstances conducive to industrial take off, including relatively high wages and cheap energy.

By the mid-eighteenth century, workers in north-west Europe had the highest standard of living of anywhere in the world. Compared with other parts of the world, these workers had higher levels of consumption of meat, dairy products and more highly refined foods like white bread. Relatively high wages were essentially a consequence of the strong economic growth of the period and helped lead to higher levels of consumption and education. Agricultural productivity had been improving such that the agricultural workforce declined, from about 70 per cent in England in 1600 to 36 per cent by 1800.[6] Changes in land ownership and the development of a more commercial and capitalist approach to agriculture helped stimulate agricultural productivity. Population growth and income growth helped increase demand for food and the burgeoning populations of towns and cities meant growing markets for agricultural produce. Economic growth in towns and cities also helped draw workers from the land. The growth in the use of horses on farms helped ensure that farm output and labour productivity improved to meet the rising demand from the towns.

The early development of the coal industry in Britain, again part-fuelled by the application of horsepower, meant that the British economy benefited from the cheapest energy in the world. The medieval economy was "propelled by animals, humans, water and wind".[7] The development of the coalfields of Northumberland and Durham helped ensure British coal production increased sixty-fold between 1560 and 1800, by which time Britain was producing most of the world's coal. Even after having been shipped to London, the price of coal from northern England was moderate by international standards. Coal was also important for the technological spin-offs from the industry, including the steam-engine and railways. British firms could pay higher wages than their French counterparts and yet remain internationally competitive because cheaper energy offset higher wage costs.

The coal industry spawned the development of the steam engine and then the railways. The steam engine was developed by Thomas Newcomen and first brought into service in 1712 to drain a coal mine in Dudley, a task that had previously been carried out using horse-powered pumps or water wheels. It was a remarkable feat of science and engineering but was only initially viable as a technology because the fuel to power it at coal mines was free and ubiquitous. By 1733, there were about 100 engines in use and by 1800 there were around 2,500.[8] Notably, while steam pumps replaced the horse for pumping work, horses were still required to transport the fuel necessary for the pumps to operate. Steam engine technology was steadily developed from the first primitive models and improvements focussed on better fuel efficiency and the pounds of coal required per horsepower-hour. In the 1760s, the work of John Smeaton and James Watt ensured a step-change improvement in the efficiency of the steam engine which gradually began to be used for purposes other than pumping water. The reduction in the cost of power meant the technology spread and between 1830 and 1870 steam power became pre-eminent in powering machinery in factories. By 1848, Britain had already harnessed the power of a million horses in its steam engines.[9]

Cotton was the "wonder industry" of the industrial revolution.[10] The industry developed out of the production of woollen and linen goods which was reliant upon independent textile workers, often based in rural areas. Before the industrial revolution, China and India were the world's biggest producers of cotton textiles and were exporting fabrics to Europe by the late seventeenth century. From the late-eighteenth century, British production began to be concentrated in factories in urban areas, especially concentrated in the growing mill towns of Lancashire.[11] Raw materials were brought in and goods shipped out by canal, on barges hauled by horses. This was a highly efficient way of transporting goods because at a steady walking speed a horse could pull approximately fifty times as much weight by canal barge as it could by road cart.

Mechanisation of production processes revolutionised the British cotton industry. Cotton had to be cleaned of debris and then carded, scraped between two hand-held cards to align the fibres. It then had to be spun into yarn which was then woven into cloth. In the mid-1760s, James Hargreaves invented the spinning jenny, followed by Richard Arkwright's development of roller spinning. Cleaning, carding and reeling all became mechanised and the cotton mill became an efficient organisation of flows of materials, power sources and production processes. This led to huge improvements in productivity such that the cost of producing yarn in 1836 was half that of 1760, with most of the gain due to savings in capital, materials and labour.[12] The cotton mill was so efficient that by the 1830s it could outcompete production anywhere else in the world. The British industry grew rapidly, and employment reached 425,000 people, accounted for 16 per cent of jobs in British manufacturing and 8 per cent of British Gross Domestic Product.[13] This was all before the development of the steam locomotive and railway. Horses were an essential part of the haulage of goods and materials that underpinned this key driver of the industrial revolution, the growth of the British economy, and the building of industrial cities such as Manchester. As we shall see below, the growth of the English cotton industry was also intricately bound up with the transatlantic slave trade so transforming international trading systems as well as domestic economic development.

Steam engines enabled the development of the railway. Steam locomotives had to be light and efficient enough to pull trains and the tubular boiler helped improve their fuel efficiency. Railways developed rapidly and on a global scale. In 1830 there were only a few dozen miles of railway in the world, mostly the line between Liverpool and Manchester. By 1840, there were over 4,500 miles and by 1850 over 23,500 miles.[14] Early railroad advocates predicted that trains would replace the need for horses. Yet instead of displacing horses, railways had the effect of generating new demand for horse-labour in Britain. The demand became so great that a Select Committee on Horses was established in 1873 to consider concerns about the capacity of the nation to supply sufficient horses to meet the increasing demand. The Committee asked a stable keeper whose business looked after 1,200 horses a year for its clients about the impact of the railways upon the carrying trade. He replied:

> I should say that it has increased it. We thought when railways first came in that we should have nothing to do, but it has not turned out so. The greater the increase of railways, the greater will be the use of the horse … because there is the work to be done to and fro. The horses have to work in connection with the railways; for every new railway you want fresh

> horses; fresh cabs to begin with; I know one cab proprietor for instance who used to keep 60 horses, who has now 120.[15]

There was a twenty-fold increase in the number of railway passengers in Britain between 1840 and 1870 and this brought increased demand for horse-drawn cabs to drop off and pick up passengers from railway terminals. In the early 1830s there were 1,265 horse-drawn cabs registered in London and this grew to 6,800 by 1863 and to around 11,000 by 1888. Each cab required two horses to keep it on the road, so the cabs alone accounted for a growth of almost 20,000 additional horses in London over this period. Rail passengers would have also made use of London's horse-drawn buses and so contributed to the growth in the need for horses to pull buses. London bus numbers more than doubled from 620 in 1839 to nearly 1,300 in 1850 and then almost trebled to 3,696 buses by 1902. Across Britain as a whole, there were approximately 25,000 horse-buses in 1890, and numbers continued to grow. To keep each bus in service required 11 horses and so there were a peak of approximately 40,000 bus horses in London and 400,000 nationally by the early twentieth century.[16]

In addition to conveying railway passengers to and from stations, horses were employed by the railway companies to help with hauling and shunting work, even after the introduction of the steam locomotive. Horses were also used for the onward delivery of goods from railway stations. Here, large independent carriers sometimes served the railway companies, especially in London. Pickford's, a removal and haulage company established in the seventeenth century, had 1,500 horses working in London by the 1870s, mainly delivering for railway companies. It owned a large stable complex in Camden Town, now the site of Camden Market.[17] Outside London, railway companies had their own sizeable horse and van fleets. By 1890, it was estimated that there were 6,000 railway horses in London, and this high level of railway-generated demand for horses would have been replicated in railway towns across the country. "The bulk of rail-borne merchandise could hardly escape being touched by horse".[18] Privately-owned horses also grew with the increase in railway travel and more than doubled in number in the second half of the nineteenth century.

Horse-power underpinned the industrial revolution, even if it is the mechanical technologies such as the steam engine, spinning jenny and the rise of coal and steam power that dominate conventional accounts of economic change over this period. Quietly neglected by economic historians, the horse had played a central and fundamental role in creating the industrial conditions than enabled technological and economic take-off, but also continued to play a key role in the work transporting goods to and from the mills and factories and goods and people to and from the railway stations of the new industrial age.

Horses and Empire

The development of British economic geography and the British state from the seventeenth to the nineteenth century has to be seen in the context of its relations with the rest of the world, as does the role of the horse in these processes. Britain's economic development over the two centuries to the early twentieth century is the story of the building of the largest empire the world had ever known. British domestic development and the amassing of its Empire are mutually intertwined and reinforcing. By the early twentieth century, British dominions, colonies, protectorates and other territories covered a quarter of the world's land area and a quarter of its people. The economic relationships, political systems and cultural values that featured in the building and maintenance of the British Empire, as well as the physical infrastructure around trade and government, helped shaped the political economy of the early twentieth century and provide the economic and technological context for the transition to the post-horse as well as post-colonial era. Eric Hobsbawm, in his three-volume world history of the long nineteenth century, characterised the 125-year period from 1789 to 1914 as having three broad phases, the *Age of Revolution* (1789–1848), the *Age of Capital* (1848–1875) and the *Age of Empire* (1875–1914).[19] The expansion of the British Empire was a process that unfolded through these three phases, but the age of revolution was particularly characterised by the technological, social and economic changes unleashed by the industrial revolution and the change in political values surrounding the French Revolution. The age of capital saw the extension of the capitalist system across the whole world. Although capitalism was still centred on Europe and the colonial powers, the US was already beginning to emerge as a new world economic power. The age of capital consisted of the "quiet but expansionist 1850s, the more turbulent 1860s, and the boom and slump of the 1870s".[20] The age of empire saw the British Empire at its peak and the establishment of a global world order. Steam and rail travel had made intercontinental and transcontinental travel a matter of weeks rather than months and the telegraph had made the communication of information around the globe achievable in a matter of hours.[21]

The rise of Empire goes hand in hand with the strengthening of the concept of Britain, what Phillipa Levine calls 'uniting the kingdom', and Linda Colley calls 'forging the nation'.[22] Internal colonisation of the British Isles by the English, in which horses and horsepower had played a vital role, was the precursor to the development of the British Empire. Wales was formally brought within English control through a 1536 Act of Union. Scotland joined through the 1707 Act of Union but was granted much more latitude than the Welsh had been and maintained its own judicial system and national church. Wars with France and Spain in the late eighteenth century provided the background to the 1800 Act of Union that granted Ireland 100 seats in the British Parliament and meant that a United Kingdom of England, Scotland, Wales and Ireland were governed as one nation from London. Linda Colley has written of how trade and Empire played a key role in forging the nation. Britain was a trading country and by the eighteenth century perhaps one family in five derived their livelihood from trading and distribution. Domestic traders and overseas merchants tended to favour a strong state to maintain good order in commerce and markets. They benefitted from economic order and imperial growth and therefore formed a loyal commercial community of support alongside the landed elite who ran government, military, and empire.[23]

The British Empire had evolved from a set of territorial relationships centred on trade and exploration. Prior to the eighteenth century, Spain and Portugal were the world's dominant colonial powers. The English and, after 1707 the British, began to establish colonies in North America and the Caribbean. From the late seventeenth century until the early nineteenth century, the Atlantic slave trade formed a significant driving force in British trade.[24] The trade was vast in scale. People from central and western Africa were captured and enslaved and transported to North America and the Caribbean. The products of slave-labour plantations such as cotton, sugar and tobacco were transported to Europe and European goods were taken to Africa to trade for slaves. Britain was the dominant slave-trading nation, and it is estimated that between the mid-1670s and 1800, over 40 per cent of slaves that crossed the Atlantic did so in British ships.[25] By the late eighteenth century, more than twenty ships a year were built in Liverpool alone to transport slaves.[26]

Horses were used in slave plantations, including in the British West Indies, both as draught animals and as a means of transport for plantation owners. Horses on plantations tended to be looked after by enslaved labour.[27] West Indian horses were either bred locally or imported from Cuba, North America and Britain. Local horses cost between £30 and £70 between the 1780s and 1830s although 'high-bred' horses from Britain or America could cost £100 or even £200, more than the cost of buying an enslaved person. Managers and overseers would ride horses in their work on the estates and horses were used to catch runaway slaves and used by militia to intimidate slaves and put down insurrections.[28] Horse-riding and travel by carriage was a feature of everyday life among affluent colonialists in the West Indies and was part of their leisure activities. Horse-racing also spread to the colonies in the last quarter of the eighteenth century and became a popular pastime among the white elite, although horses were usually ridden in races by black jockeys. Enslaved people were not permitted to keep horses although they would work with them to power mills and cart loads. They would also often be required to walk alongside a master on a horse, often holding the horse's tail in a reflection of white power and privilege. David Lambert explains:

> "Horses and other animals were an integral part of British and West Indian life at the time, of course, but in slave societies, where some human beings performed roles that animals would elsewhere, where some human beings were treated in ways similar to animals and where animalistic discourse was common, the place of animals was especially significant. The horse was a form of social capital in West Indian societies and a symbol of (white) mastery as well as a means of transportation".[29]

After the loss of the American colonies after 1783, the era of what has been called the First British Empire came to a close and attention turned more towards India and Australasia. After victory over France in the Napoleonic wars, Britain was left the world's dominant imperial power for the century 1815 to 1914. British territories in Asia and the Pacific had accumulated through exploration and the establishment of strategic ports, trading posts and penal colonies. Botany Bay was established as a penal colony in 1786, for example. Horses arrived in Australia with the early fleets to carry out farm work, although not in large numbers. They spread rapidly and by 1800 the first feral horses (known as brumbies) had escaped from farms and were found across the continent by the 1840s.[30] By the twenty-first century there were estimated to be 400,000 feral horses across Australia, posing complex environmental management problems and public controversy around culling.[31]

In the latter period of the British Empire, the development of steamships and telegraph helped provide a technological basis for managing imperial lands and seas. Until this period, though, horses played their part in underpinning the establishment and maintenance of empire. Between 1650 and 1750, more than two hundred Eastern or Oriental horses were imported into England. Usually referred to as Arabian, their bloodlines were seen as superior, and they became important breeding stock. Within a generation, the progeny of these horses were being called 'English thoroughbreds' and their quality became important in distinguishing English or British stock from other European horses. Britain began to gain confidence when it compared itself to the Eastern empires. Donna Landry writes of how the new types of horses, and the new disposition of the body on horseback, came to signify a British aristocratic and gentlemanly style. Equestrian culture, including hunting and racing, and equine portraiture and art, "served imaginatively to express Britain's 'gentlemanly capitalist' version of mercantilism during the nation's rise to global economic importance between the late sixteenth century and mid-nineteenth century".[32] Landry explains:

> The arrival of Turkish, Barb and Arabian horses revolutionised not only English racing and equestrian culture but early modern culture more broadly .... Along with the horses came Eastern ideas about horsemanship, about the relation between horses and humans, and about forms of representation of this relationship.[33]

An equestrian Anglophilia developed, where distinctive British styles and approaches to horsemanship and horse culture became adopted elsewhere around the world. In France, a vogue developed for British horses, grooms and equestrian vocabulary. Horses were integral to how national identity and culture were projected abroad and consumed.

Horses were also used as an instrument of colonial oppression and power. Not only would they carry colonists from one place to another, but they could also be used to exert power over the colonised.[34] Their introduction or reintroduction to colonial territories disrupted existing ecosystems and cultures. Christopher Columbus took horses to the Caribbean on his second voyage in 1493 and from there the species was re-introduced to North America.[35] The sight of a man riding on a large beast was reported to have terrified the indigenous people of the Caribbean. French settlement in Quebec and English and Dutch colonies along the American east coast meant that by the mid-seventeenth century horses were well established throughout Lower Canada, New England and Virginia and Pennsylvania. As animals were left to forage for themselves, so they spread west across North America. Large herds developed and horses were re-domesticated by the indigenous American Indians and used to radically change the ways that bison were hunted. From the 1750s to the 1870s, North American Plains Indians made full use of the horse. The equestrian culture and horsemanship among the Plains Indians, particularly among the Lakota people, was an important part of the popular iconography around their ability to resist American military forces.[36]

Horses had become the main draught animal for American agriculture by the end of the eighteenth century and helped significantly improve agricultural labour productivity during the nineteenth century. The number of horses on American farms continued to increase in the latter part of the nineteenth century, rising from 7.3 million in 1870 to 17.8 million by 1900.[37] American horsepower was also being used in industrial processes and to haul freight and passengers in American cities, with new horse-car railroads opening in New York, Boston and Philadelphia from the 1830s. As a result, there was growing demand for larger draught horses which seemed to bring "almost limitless opportunities for British and European breeders".[38] Shire stallions began to be exported to Canada, followed by Clydesdales and by 1870, more than seventy Clydesdale stallions were covering mares in the colony.[39] The flow of emigrants from Europe boosted the US population and the demand for horses. Between 1900 and 1910, the human population rose by 21 per cent, while horses and mules increased by 70 per cent from 13 to 23 millions. Imports of Clydesdales to the US and Canada increased from 360 in the period 1892–96 to 5,584 in the period 1907–11.[40]

Horses and the Economic Geography of Nineteenth Century Britain

The nineteenth century saw a surge in the numbers of horses in Britain. The point of 'peak horse' came in the early years of the twentieth century but by that stage the technologies and processes were in place that would ensure that horse numbers declined as the transition to the post-horse age played out. By the end of the nineteenth century, the horse economy was at its most advanced and specialised. Firms even existed to manufacture straw bonnets for horses to keep the sun, and flies, out of their faces. As historian F.M.L. Thompson put it, writing of the turn of the twentieth century, "the horse-drawn world reached the apogee of its refinement and specialisation at the very moment when motor vehicles were becoming sufficiently numerous for businessmen to notice their existence".[41] It was the petrol engine rather than the steam engine that undermined the cost-effectiveness of the horse and eventually led to its replacement.

Horse numbers in Britain trebled during the nineteenth century, with most of the increase coming in the second half of the century through growth in commercial horses to pull goods and passengers in the growing towns and cities. Britain had become a much more urbanised nation during the century and by 1901 about 80 per cent of the population lived in towns. Numbers of horses for urban and industrial uses grew much more rapidly than those for agriculture and so their proportion of the total grew from a third in 1811 to a half by 1901.[42] British growth in horse numbers was marked, but not the highest in the world. In the US, for example, the horse population grew from 6.8 million in 1867 to about 18 million in 1900.[43]

Understanding British horse numbers in the nineteenth century is an inexact science. Horses in agriculture could be tracked through agricultural census returns, but there was not a full census of all horses until well into the twentieth century. Even agricultural returns have some problems. They collected information from all agricultural holdings, but this included small holdings of land attached to premises such as butchers or inns, and not all horses on agricultural holdings were farm horses working in agriculture. In 1811, around a fifth of horses in Britain (251,000) were in the commercial sector and a further fifth (236,000) were private horses pulling carriages. Almost two fifths of all horses were not on farms, while more than three fifths were agricultural horses. By 1871, the total number of horses in Britain had increased from around 1.3 million to 2.1 million, an increase of almost two-thirds since 1811. However, the balance between commercial horses, private horses and agricultural horses remained the same as sixty years earlier. By the end of the century, total horse numbers were 154 per cent higher than in 1811. However, by then the balance had shifted away from horses on farms and towards commercial horses, with over a third of all horses now in the commercial sector and under a half on farms.[44]

Major changes in industrial structure and transport infrastructure account for the marked growth in horse numbers, especially in the latter part of the nineteenth century, as well as the changes in the use of horses. The production of horses was principally still from farms, and usually horse-breeding was an offshoot of a farm business, rather than a wholly specialised business. Some horses were imported, but imports and exports of horses of working age generally balanced out by the end of the century. Draught horses came from Denmark, Holland, Belgium and France. Ponies were imported from Scandinavian countries, East Prussia and Poland. Riding and driving horses came from Hungary and Hanover. Some horses were traded across the Atlantic and a few were imported from the Russian steppes and from Latin America. There was a steady supply from southern Ireland, where the horses had a good reputation as hunters and saddle horses and almost 40,000 horses a year were coming from Ireland by the end of the century. Older horses at the end of their working lives were exported to Belgium and the Netherlands for slaughter for meat. These exports ran at 29,000 a year in the late 1890s but grew to 51,000 by 1906.[45]

The point of ‘peak horse’ also meant ‘peak horse feed’ and ‘peak horse dung’. Each town horse produced three to four tons of dung each year.

Properly mixed with straw, this produced 12 tons of good manure per horse per year but if it was left on the street, it was a mounting problem.[46] Town horses required an estimated 1.4 tons of oats or corn and 2.4 tons of hay per head per year. Taking the town horses alone, this required 1.5 million to 2.0 million tons of oats and 2.5 million to 3.5 million tons of hay by the end of the nineteenth century, equivalent to the entire British output of oats and one third of its hay production.[47] In practice, much horse feed was imported, including maize, but feeding Britain's 3 million horses at the turn of the twentieth century was a major force driving agricultural land use patterns. Thompson calculated that by the end of the nineteenth century, it was costing 10s a week to feed a town horse, suggesting a total annual feed bill of between £26 million and £39 million, meaning that the cost of feeding horses was almost equivalent to the annual value of all crops sold off British farms.[48] At the turn of the twentieth century, British agriculture was in its third decade of the long agricultural depression as Britain sourced cheap food from around the world and farmers were struggling at the mercy of world market prices. Feeding horses was one of their most important markets of the day. The area of oats grown in Britain rose from 1.1 million hectares (2.7m acres) in 1875 to 1.5 million hectares (3.7m acres) by 1919. The proportion of the area of total cereals crops made up by oats rose from 31 per cent to 48 per cent over the same period.[49]

Farm horses were less likely to be fed hay until the spring when their workload increased. In contrast, town horses needed hay virtually all year round. There were 31 towns and cities in Britain estimated to have at least 3,000 town horses by the end of the nineteenth century which, with London, accounted for the bulk of the fodder trade. Britain's horses were more geographically concentrated in London than humans were. Because hay was bulky, it tended only to be transported short distances, and rarely more than 30 miles, although London's hay catchment was extended as a result of the canal network. Even after the development of the railways, most of London's hay came from the Home Counties. Hay was usually grown by farmers who had other business interests, as it took so little time within the annual calendar. Hay would be carted in one direction and horse dung in the other. Dung was also transported by rail. In the early 1860s, one farm of 180 hectares (445 acres) in Hertfordshire was receiving 1,000 tons of horse dung a year, delivered by rail.[50]

By the mid-nineteenth century, London's horses were consuming around 200,000 tons of hay a year, and the hay farms of the Home Counties were meeting between a half and two-thirds of this. The rest came from further afield by canal or rail. Similarly, around Manchester, Liverpool and Birmingham, large numbers of town horses sustained a healthy market for commercial hay growers in the surrounding regions. As town horse numbers grew in the latter half of the nineteenth century, so the share of hay supplied by specialist hay growers fell. Improvements in haymaking, including the greater use of horse-drawn mowers, improved the quality of hay produced by a wider and non-specialist farming population, and more hay was drawn into cities from further afield. From the late 1870s, hay began to be pressed to take up less space and the practice became more widespread by the 1890s. Hay presses could reduce hay to at least one-fifth of its original bulk and so saved transport costs. Between 1888 and 1899, 44 different manufacturers exhibited hay pressing machines at Royal Shows.[51]

The growth in horse numbers during the latter decades of the nineteenth century caused problems. The nuisance posed by horse manure had become a public controversy in New York during the 1880s and was picked up in London too.[52] The growth trends in horse numbers and the volume of dung produced were extrapolated forwards to raise concern about the possible accumulating problem of managing and disposing of dung. Advocates for motor vehicles began to campaign highlighting the problems of the horse. Sustaining the growth of town horses had led to a self-perpetuating spiral of problems around space, congestion and noxious dung. One engineer and founding member of the Royal Automobile Club made the case in a presentation to engineers in London in 1898. He said:

> We are rapidly reaching a density of traffic impossible to contend with [in London] if the streets are to be anything more than lines of free railways on which every cab and omnibus and coal wagon driver is a traffic manager … As early as possible motor cabs and omnibuses must displace horses so as to save space now occupied by them, avoid the spreading all over the streets of that which renders them dirty and unhealthy, and avoid the pounding of pieces of wood, asphalte and others of the best paving with three-cwt hammers of iron-shod horses' feet … These horses are not only destroying our streets and roads at double the rate at which even the iron-shod wheels would do it, but they are themselves the origin of an enormous quantity of street haulage. Nearly all the food and fodder has to be hauled over some parts of the roads and streets … The street refuse, which employs an army of scavenging boys, and then scavenging carts and more horses, to carry away from the street boxes that which is not blown into the air we breathe or washed into the sewers, will represent another enormous quantity of street carting … At least one-third of the space now occupied by the horses of the vehicles from great railway vans to hansom cabs, can be saved by the employment of motor vehicles.[53]

Campaigners in Britain pointed to the experience in the US where the growth of horse numbers in cities had also been marked and was beginning to cause controversy. Notable in these campaigns was the struggle over urban space. While the advent of steam power had unexpectedly led to greater demands for horses and their work, the advance of the electric tram, petrol engine and motor vehicle ensured that Britain moved through its 'peak horse' moment. However, the process was more protracted than may have been anticipated and is commonly represented in academic and popular accounts of technological change.[54] During the twentieth century, the horse became redefined as much more a creature of sport and leisure, with working horses gradually retreating to only a few specialised roles.

The Stuttering Decline of the Horse

The late nineteenth century had seen horse numbers rise but it was, as Theodore Barker put it, "something in the nature of an impressive final gallop".[55] 'Peak horse' was reached in Britain probably during the first decade of the twentieth century with a population of almost 3.3 million animals. After 1910, horse numbers in towns declined so dramatically that they almost disappeared. Horses did hold on in agriculture and coalmining for a few decades more. On farms, they still provided more horsepower than tractors in 1939 and during the Second World War, the County War Agricultural Executive Committees that steered farm production encouraged farmers not to use scarce tractors for tasks when a horse would do.[56] There were still more horses than tractors in 1950 but, as we shall see in Chapter 5, tractors eventually took over in farms and fields too. The decline in horse numbers had dramatic implications for the horse economy, which included a whole range of industries that depended on the horse. The decline also brought changes in the pattern of demand for agricultural products, as no longer were such quantities of feed required for horses. Horse-breeders and knackers, blacksmiths and farriers and the manufacturers of horse-drawn vehicles all saw their markets shrink, and the livelihoods of those who drove and cared for horses and repaired and maintained their paraphernalia melted away. Falling horse numbers and falling demand for horse feed contributed to the long economic malaise in British farming through the 1920s and 1930s. The transition from the age of the horse to the post-horse age was one that attracted little attention from historians and social scientists for many years. The emphasis of analysis tended to be on the spectacular spread of the new technologies, rather than the implications for those who were dependant on the old order. A exception is the work of economic historian F.M.L. Thompson who gave his inaugural lecture at the University of London on the topic of *Victorian England - the Horse Drawn Society* in 1970 and went on to produce several valuable contributions to the economic history of the horse.[57] Britain was not alone in this transition, and the spread of motor vehicles and tractors meant a marked decline in horse numbers across the world over the first half of the twentieth century.

In the US, horse-drawn tramways on rails had become the mainstay of urban public transport systems in cities. An electric-powered overhead trolley system was developed in Richmond, Virginia in 1888 and the innovation spread rapidly. Electrification changed the economics of the tramways industry and given people's tolerance of commuting times was relatively constant, the greater speeds achievable by electric tramways enabled the physical expansion of American cities as electric streetcars replaced the horse. By 1890, one fifth of America's total mileage of urban tramways was electrified and by 1895, there were over 12,000 miles of tramway, with only 1,230 miles of the old horse-drawn lines remaining.[58] At its peak, about 240,000 horses were employed on the US urban streetcar system, but by 1902, there were fewer than 9,000 horses left doing streetcar work, a spectacular collapse in horse numbers.[59] The total working horse population in the US at the turn of the twentieth century was almost 18.4 million, the vast majority of which were, by then, working on farms.[60] The number of farm horses in the US did not initially drop with the early growth in tractor numbers. Instead, new tractors were used to expand power and production. Between 1900 and 1910, farmers brought tractors but did not substitute them for horses. After 1910, the numbers of working horses on American farms did decline from 17.4 million to 13.5 million in 1930, 10 million in 1940, and 3 million in 1960. In horsepower terms, American farmers owned 4.6 million tractors and almost 3 million motor trucks by the 1960s, and the contribution of horsepower by horses and mules fell from a peak of 22.4 million hp in 1920 to 12.5 million in 1940 and 1.25 million by 1969.[61]

British tramways were electrified around 1900, although London lagged other cities. However, London led in the mechanisation of omnibuses. A technical challenge had been to cope with the starting and stopping at stops and in heavy traffic. Initially horses were better than the early motor engines, but once engines improved, change came quickly. In 1903 London had 3,623 horse-buses and just 13 motor buses, but just ten years later the proportions had been reversed and only 142 horse-buses were left alongside 3,522 motor buses.[62] Horse buses disappeared altogether soon afterwards. Motor taxi cabs replaced horse-drawn hackney carriages and most private owners of horse-drawn carts and carriages switched to motor vehicles in the first decade of the century. By 1911, a traffic census in London found that only 13 per cent of passenger vehicles were stilled pulled by horse and two years later this proportion had fallen to 6 per cent. "In the world's largest city the horse had been replaced for passenger purposes within the space of little more than a decade".[63] A similar transformation took place in Paris and other major European cities.

The First World War was also a pivotal moment of transition in the use of horses in warfare. All the major forces began the War with cavalry units that had previously been considered important offensive units in war. However, their use declined as defensive machine guns and artillery fire reduced their effectiveness. Military horses were still used for logistical support behind the lines and the volume of military stores and rations required to support the troops was such that there was a greater need for horses in this logistical role than there had been in earlier conflicts. A huge effort had to be undertaken to ensure sufficient horses to support the armed forces and horses were sourced from far and wide. The War created an international equine economy and disrupting the enemy's supply of horses became a feature of the military struggle.[64] At its outbreak in 1914, the British army had approximately 25,000 horses. By mid-1917, it had almost 600,000.[65] In an analysis of their role in the conflict, John Singleton suggests that horses were "as indispensable to the war effort as machine guns, dreadnoughts, railways and heavy artillery".[66] It has been estimated that the British forces lost almost 500,000 horses during the war in total. An obdurate and petulant approach to the use of cavalry forces in the War failed to adapt to the changes in warfare wrought by the new technologies. The results were disastrous for soldiers and horses alike. An officer of the Highland Light Infantry wrote of the battle of Arras in April 1917:

> An excited shout was raised that our cavalry was coming up. Sure enough, away behind us, moving quickly in extended order down the slope … was line upon line of mounted men, covering the whole extent of the hill-side as far as we could see … It may have been a fine sight, but it was a wicked waste of men and horses, for the enemy immediately opened up on them a hurricane of every kind of missile he had … They bunched behind Monchy in a big mass into which the Boche continued to put high explosive, shrapnel, whizz-bangs, and a hail of bullets … The horses seem to have suffered most, and for a while we put bullets into poor brutes that were aimlessly limping about on three legs, or else careering about madly in their agony; like one I saw that had the whole of its muzzle blown away.[67]

Between the two world wars, the development of motorised armoured forces meant the military use of horses declined markedly, although many in the military were reluctant to see the horses go. The Secretary of State for War, Duff Cooper, reported to Parliament on the progress with the mechanisation of mounted regiments in 1936. He praised the cavalry on the way they had accepted the changes: "all the traditions of the regiment are bound up with their horses. It is like asking a great musical performer to throw away his violin and devote himself in future to a gramophone. It is a great sacrifice for the cavalrymen".[68] By the time of the Second World War, the British Army still had around 6,500 horses but mounted cavalry regiments were replaced by the Royal Tank Regiment and the Royal Armoured Corps. Although some mounted regiments were used by Soviet forces on the eastern front during the Second World War, their use soon petered out.

The use of motor vehicles spread around the world in the early twentieth century. They suited not just passenger traffic but also the delivery of goods, especially higher-value freight-postal services and commercial delivery. Although early adoption was hampered by mechanical unreliability and frequent breakdowns, as the design and manufacture of motor vehicles improved so the switch from horse to motor in transporting passengers and high value freight was rapid. For heavier and bulkier loads, steam engines continued to be preferred.[69] The change in transporting goods traffic took longer than changes to trams and omnibuses. In 1913, when there was barely any hauling of passengers by horse in London, almost 90 per cent of goods vehicles in the city were still horse-drawn. The switch first came with higher value freight where speed was a significant advantage. Deliveries of goods from shops and furniture removals were among the first services to motorise. Other sectors followed only slowly as motor vehicles became more readily available and more efficient and cost-effective. Almost a decade after horse-drawn passenger transport had disappeared from

London's streets, there were still 237,000 horse-drawn vehicles in Britain principally hauling lower value goods. This figure dropped during the 1920s to 66,000 in 1929 and just 12,000 by 1937. Railway companies continued to use horse-drawn wagons for delivery work.[70] In 1934, the four main line railway companies still employed almost 15,000 horses,[71] and a few continued to be used for a few more years, even after the Second World War. After that, horse-drawn vehicles largely disappeared from Britain's streets, except for a few used by breweries, the odd rag-and-bone man, coal merchant, or funeral director.

Working horses persisted in coalmining. Mechanisation was relatively slow in the first half of the twentieth century and when the pits were nationalised after the Second World War, the National Coal Board inherited 21,000 pit ponies still at work in the mines. Although mechanisation and the closure of mines meant the number of pit ponies diminished, there were still some at work up to the end of the twentieth century. For example, the last horse, Flax, finished its shift at Ellington Colliery, ending three hundred years of horse-power in Britain's mines, on 24th February 1994 and the last working pit pony to survive in Britain, Tony, died in retirement in 2011 aged 40.[72]

In contrast to the US, where total horse numbers were dominated by those working in agriculture, in the UK agriculture accounted for only around a third of all horses in the first decade of the twentieth century. The decline in town horses therefore made a more significant impact on total horse numbers. There were around 1.1 million horses working in British agriculture in 1910, although that number fell to 960,000 by the end of the First World War.[73] The total number of horses on farms increased after the War, but the numbers used for agricultural work fell away during the 1920s and 1930s. The decline in total horse numbers was slower in France than in Britain. Although town horse numbers declined over a similar period, a larger farm horse population and slower adoption of tractors meant French farm horse numbers held up during the 1920s and early 1930s. In 1934, there were still 1.1 million horses and mules on French farms, similar to the British peak level more than twenty years earlier. Even after the Second World War, the French farm horse population fell much more gradually than had been the case in the US and Britain. Non-military horse numbers continued to rise in Germany into the 1920s and in 1939 it was three times the French population and five times the British. The picture in Britain and Europe is therefore of a prolonged persistence of horses in agriculture well into the twentieth century and long after they disappeared from city streets and from much of industry (except coalmining). However, the pattern of transitioning to a 'post horse' world was geographically variable and influenced by national agricultural contexts.

Conclusions

Horses played a crucial role in the evolution of the industrial revolution and the rise of the British Empire. Although conventional accounts of the industrial revolution tend to focus on the transformational role of new technologies such as the steam engine, it was the horse that was vital in hauling the coal, by cart and by barge, to the steam engines in the first place. Early steam power was crucially underpinned by horsepower. The effects of the industrial revolution and the rise of Empire transformed the place of Britain in the world and brought a new world order centred on British trade and imperial relations. Through the height of British imperial power during the nineteenth century, the number of horses grew significantly.

The biological constitution of British horse stock developed over the nineteenth century with significant implications. Eastern bloodstock transformed the notion of what a horse was in Britain and helped change the relationship between horses and people. As Donna Landry put it, "relations of economic necessity and traditional usage became overlaid with new kinds of pleasure and poetry".[74] Imperial pride developed in step with national pride at the improving horse stock. Dramatic growth in horse numbers in the second half of the nineteenth century was, perhaps paradoxically, fuelled by the development of the railways and the greater demands for personal mobility and the movement of goods and materials that were generated by economic growth. Britain's 'peak horse' moment coincided with the height of the British Empire. There was a sense of confidence in the invulnerability of the horse at the turn of the twentieth century,[75] but the new 'post-horse' technologies of the motor car, motor truck, omnibus, electric tram and tractor were all poised to play their part in the variegated transition to the post-horse economy. Each contributed to the reduction of horse numbers, although working horses were remarkable persistent in some parts of the economy well into the twentieth century, especially in agriculture.

Notes

1 Hobsbawm, E. (1962) *The Age of Revolution 1789–1848*. London: Weidenfeld and Nicholson, p.43.
2 Hobsbawm (1962), p.44.
3 Williams, E. (2022) *Capitalism and Slavery*. London: Penguin, p.120.
4 Williams (2022), p.121.
5 Hobsbawm (1962), p.60.
6 Overton, M. (1996) *Agricultural Revolution in England: The Transformation of the Agrarian Economy 1500–1850*. Cambridge University Press, p.82.
7 Allen, R. (2009) *The British Industrial Revolution in Global Perspective*. Cambridge: Cambridge University Press, p.81.
8 Allen (2009), p.162.
9 Hobsbawm (1962), p.68.
10 Allen (2009), p.182.
11 Phelps, A. *et al.* (2018) *The Textile Mills of Lancashire: The Legacy*. Lancaster: Oxford Archaeology North.
12 Allen (2009), p.184.
13 Allen (2009), p.182.
14 Hobsbawm (1962), p.61.
15 quoted in Thompson, F.M.L. (1976) Nineteenth century horse sense, *Economic History Review* 29, 60–81, p.65.
16 Thompson (1976), p.65.

17 Edgerton, D. (2019) *The Shock of the Old: Technology and Global History since 1900*. (Second edition) London: Profile, p.33.
18 Thompson (1976), p.66.
19 Hobsbawm (1962); Hobsbawm, E. (1975) *The Age of Capital 1848–1875*. London: Weidenfeld and Nicholson; Hobsbawm, E. (1987) *The Age of Empire 1875–1914*. London: Weidenfeld and Nicholson.
20 Hobsbawm (1975), p.10.
21 Hobsbawm (1987), pp.13–4.
22 Levine, P. (2013) *The British Empire: Sunrise to Sunset*. (Second Edition). London: Routledge; Colley, L. (1992) *Britons: Forging the Nation 1707–1837*. London: Pimlico.
23 Colley (1992).
24 Williams (2022); see also Thomas, H. (1997) *The Slave Trade: The Story of the Atlantic Slave Trade: 1440–1870*. London: Simon & Schuster.
25 Levine (2013), p.21.
26 Levine (2013), p.19.
27 Lambert, D. (2015) Master – Horse – Slave: Mobility, Race and Power in the British West Indies, c.1780–1838, *Slavery & Abolition* 36, 618–641.
28 Lambert (2015).
29 Lambert (2015).
30 Mitchell, P. (2015) *Horse Nations: The Worldwide Impact of the Horse on Indigenous Societies Post-1492*. Oxford: Oxford University Press, p.328.
31 Symanski, R. (1994) Contested realities: Feral horses in Outback Australia, *Annuls of the Association of American Geographers* 84, 251–69; Department of Sustainability, Environment, Water, Population and Communities [Australian Government] (2011) *Feral Horse (Equus Caballus) and Feral Donkey (Equus Asinus)*. Canberra: Department of Sustainability, Environment, Water, Population and Communities.
32 Landry, D. (2009) *Noble Brutes: How Eastern Horses Transformed English Culture*. Baltimore: Johns Hopkins Press, p.3.
33 Landry (2009), p.11.
34 Anderson, V. (2004) *Creatures of Empire: How Domestic Animals Transformed Early America*. Oxford: Oxford University Press, p.83.
35 Mitchell (2015).
36 Hämäläinen, P. (2003) The rise and fall of plains Indian horse culture, *Journal of American History* 90, 833–62, p.833.
37 Moore-Colyer, R. (2000) Aspects of the trade in British pedigree draught horses with the United States and Canada, c.1850–1920, *Agricultural History Review* 48, 42–59, p.44.
38 Moore-Colyer (2000), p.45.
39 Moore-Colyer (2000), p.46.
40 Moore-Colyer (2000), p.54.
41 Thompson (1976), p.60.
42 Barker, T. (1983) The delayed decline of the horse in the Twentieth Century, pp.101–12 in F.M.L. Thompson (ed.) *Horses in European Economic History: A Preliminary Canter*. Reading: British Agricultural History Society; see also Thompson, F.M.L. (1976).
43 Barker (1983), p.102.
44 Thompson (1976), p.80.
45 Thompson (1976), p.76.
46 Thompson (1976), p.77.
47 Thompson, F.M.L. (1983b) Horses and hay in Britain 1830 to 1900, pp.50–72 in F.M.L. Thompson (ed.) *Horses in European Economic History: A Preliminary Canter*. Reading: British Agricultural History Society, p.60.
48 Thompson (1976), p.78.

49 Marks, H. and Britton, D. (1989) *A Hundred Years of British Food and Farming – A Statistical Survey*. London: Taylor and Francis, p.161.
50 Thompson (1976), p.65.
51 Thompson (1983b), p.68.
52 McShane, C. and Tarr, J. (2007) *The Horse in the City: Living Machines in the Nineteenth Century*. Baltimore: Johns Hopkins Press, pp.26–7.
53 quoted in Barker (1983), p.104.
54 Edgerton (2019).
55 Barker (1983), p.101.
56 Forrest, S. (2016) *The Age of the Horse: An Equine Journey through Human History*. London: Atlantic, p.181.
57 Thompson, F.M.L. (1970) *Victorian England: The Horse-Drawn Society*; London: Bedford College; see also Thompson, F.M.L. (ed.) (1983) *Horses in European Economic History: A Preliminary Canter*. Reading: British Agricultural History Society.
58 Barker (1983), p.105; see also McShane and Tarr (2007).
59 McShane and Tarr (2007), p.172.
60 Barker (1983), p.106.
61 Barker (1983), p.107.
62 Thompson (1976), p.61.
63 Barker (1983), p.108.
64 Singleton, J. (1993) Britain's military use of horses 1914–18, *Past & Present* 139, 178–204, p.202.
65 Singleton (1993), p.178.
66 Singleton (1993), p.178.
67 Quoted in Ellis, J. (2004) *Cavalry: The History of Mounted Warfare*. Barnsley: Pen and Sword Books, p.176.
68 Quoted in Ellis (2004), p.182.
69 Brown, J. (2008) *Steam on the Farm: A History of Agricultural Steam Engines 1800 to 1950*. Marlborough: Crowood Press.
70 Barker (1983), pp.108–9.
71 Thompson (1976), p.66.
72 Paxman, J. (2021) *Black Gold: The History of How Coal Made Britain*. London: William Collins, p.18; Kirkup, M. (2016) *Pit Ponies*. Newcastle upon Tyne: Summerhill Books, p.44.
73 Marks and Britton (1989), p.18.
74 Landry (2009), p.5.
75 Thompson (1976), p.62.

4 Horses and the Town

Introduction

In considering the role of the horse in the evolution of different types of places, it can be difficult to separate the country from the town because the two were not only interconnected but increasingly interconnected over time. In the country, horses played a vital role in extending personal mobility and the development of agricultural practices and were an integral part of country leisure pursuits such as hunting. They also played a role in the development of industrial production, which took place in both town and country, in growing industries such as tanneries and brewing. Like most of Europe, Britain became more urbanised during the seventeenth, eighteenth and nineteenth centuries. However, urbanisation was much more marked in Britain and as industrialisation progressed at a faster pace, so did the urbanisation that accompanied it. England's urban population, defined as those living in settlements of 10,000 people or more, was estimated at 8.8 per cent of the total population in 1650, so ranking as the eighth most urbanised of sixteen territories in Europe. By 1700 the proportion had risen to 13.3 per cent (and the rank to fifth), 16.7 per cent in 1750 (and ranked third) and 20.3 per cent in 1800 (second only to the Netherlands). By 1850 England's urban population was ranked first at over 40 per cent, with Scotland second at 32 per cent.[1]

The role of the horse was fundamental to early industrialisation, the growth of mobility of people and goods and the level of interconnectedness between people and places. The growth of horse use underpinned greater personal mobility both between urban centres and also within them. At the same time, these linked processes of increased mobility, industrial development and urbanisation, together served to radically change the settlement structure and urban geography of the nation. The active role of horsepower and horse-enabled human mobility in these processes has not been the subject of much analysis by geographers or by social scientists more generally. The economic and population geography of Britain's towns and cities and British historical and social geography have helped trace the evolution of the nation, but studies have not tended to consider the human-horse relationship within processes of

DOI: 10.4324/9781003454359-4

change. Although the horse was highly instrumental in Britain's marked and distinctive urbanisation, its contribution has been left neglected.

In the United States in recent years, research by historians has shed new light on the urban history of the horse. A notable contribution is that of Clay McShane and Joel Tarr which examined the role of the horse in the development of the nineteenth century American city, which particularly focuses on the conception of the horse as a 'living machine'.[2] At the same time, work by Ann Norton Greene demonstrated the pivotal role of the horse in the rise of American industrial power and the development of the landscape.[3] This chapter places the horse at the heart of an urban historical geography of Britain. It first examines the role of the horse in the radical enhancement of mobility and the flow of people and goods around Britain, especially from the sixteenth to the eighteenth centuries. The chapter examines the ways that geographies of industrialisation, urbanisation and deepening interconnectedness all interrelated, but around the central relationship between people and horses. It goes on to explore in more detail the roles of the horse in the worlds of work and of play in the eighteenth-century city before considering the implications for the shape of cities. It concludes by briefly reflecting on the de-horsification of the city in the twentieth century.

Horses, Mobilities and Flows Between Towns

Before the horse, people were limited to travelling over land on foot. The horse expanded the territory over which people could travel and facilitated greater interaction between settlements. Horses also stimulated human mobility within the larger settlements such as London. Early histories of the development of roads and overland travel have suffered in comparison to travel overseas and between countries because of a lack of the systematic documentary evidence provided by customs records at ports. Transporting goods was more expensive by land than by water, and so the coast and navigable rivers were the main transport routes of the early medieval period. The term 'road' was rarely used before the seventeenth century, and people travelled along pathways, port-ways, trade-ways, market-ways, church-ways, drift-ways, ride-ways, sandy-ways and green-ways. The Romans had left a well-engineered system of long-distance streets although they were not well-maintained after the Roman period. As volumes of traffic by horse grew in the fifteenth and sixteenth centuries, so the poor state of the main highways became a source of concern. Steps were eventually introduced in the seventeenth century to restrict the size of draft animal teams, the weight of freight and the width of carts and wagons' wheels to reduce damage to transport infrastructure.[4]

It has been estimated that there were around 500,000 horses in England by the early fourteenth century.[5] The turnpike system first developed around this time to collect tolls to pay for paving and repairs to roadways. The use of pavage grants expanded during the fourteenth century and some 435 different pavage grants were awarded by the Crown between 1247 and 1477.[6] Grants tended to be largely for roads and streets within towns rather than between towns. As travel on horseback and by horse-drawn wheeled vehicles grew in the early modern period, legislation developed to formalise the responsibilities for maintenance of the highways. Highways and bridge statutes were introduced to place responsibilities on towns or local landowners to ensure repair and maintenance of roads and bridges from at least the 1520s. Compliance was patchy. From the 1660s, the turnpike principle began to be applied more extensively where travellers were charged to use highways.

In the early sixteenth century, the calculation of time and distance were not standardised and what counted as a mile or an hour varied across different parts of the country. A standardised time for the whole country is a relatively recent development. For example, Plymouth's clocks ran more than sixteen minutes behind those in London.[7] Given that travel times by foot or by horse were relatively slow, this did not matter much. Most travel was on foot in the early modern period. The cost of keeping a horse would have been more than the average wage for a rural worker, although horse ownership did grow among the wealthier classes, especially between 1600 and 1700, in what Mark Brayshay calls "a remarkable democratisation in horse ownership".[8] Will records show that for those with personal estates valued at £100 or more, horse ownership was common after the 1570s. Four-wheeled carts were introduced from the Netherlands and Belgium in the mid-sixteenth century and began to carry both passengers and freight between London and major towns of the time such as Canterbury, Norwich, Ipswich and Gloucester.[9] This development only served to increase wear and tear on the roads. Subsequently, for the wealthier, coaches began to be used to carry small groups of people. The coach was suspended on leather straps but even on the best roads it would have been a bumpy ride and they tended initially to be used within towns rather than between them. Traveller and writer, Fynes Moryson, described arrangements in 1617.

> Coaches are not to be hired anywhere but only at London; and howsoever England is for the most part plaine, or consisting of little pleasant hilles, yet the waies farre from London are so durty, hired Coachmen doe not ordinarily take any long journeys, but onely for one or two daies any way from London, the ways so farre being sandy and very faire, and continually kept so by labour of hands. And for a days journey, a Coach with two horses used to be let for some ten shillings the day (or the way being short for some eight shillings, so as the passengers paid for the horses meat) or some fifteen shillings a day for three horses, the Coach-man paying for his horses meate.[10]

In the sixteenth century, the speed of delivering messages by horse improved significantly because of improvements in infrastructure and in the system of swapping horses. In 1600, private individuals could send a letter and expect it to travel by horse at a rate of about 100 miles a day, but the fastest messages with well-managed switches to fresh horses could achieve as much as 200 miles a day. In 1603, news of the death of Elizabeth I reached Scotland in 62 hours, twice as fast as the news of the death of Edward I had reached London from Scotland a century earlier.[11]

By the 1620s, horse-drawn wheeled vehicles were seriously challenging the long supremacy of travelling by foot or on horseback. By the time of William III's reign (1689–1702), the amount of movement and mobility among England's six million inhabitants meant towns and cities were increasingly being complained about as congested. Brayshay's study reveals how early modern society was strikingly mobile, especially because of the horse.

> Travel was commonplace; it was certainly not regarded as unusual. Journeys undertaken for mercantile purposes were almost certainly the most numerous; those of carriers, peddlars, merchants and their agents, and traders going to local markets and fairs. However, the mobility of craftsmen and the movement of materials also occurred on a substantial scale.[12]

Horses enabled more extensive mobility and meant more frequent encounters between people from further afield. This greater mobility became integral to the establishment and maintenance of political and economic power and the crystallisation of nation-states across Europe was a critical development of the early modern period. Greater mobility and interaction also underpinned a series of social and cultural changes which helped reinforce the idea of nationhood.[13] In summing up a nation strikingly on the move in Tudor and Stuart times, Brayshay writes:

> Carriers regularly freighted goods by cart, wagon and packhorse and plied a network that grew wider and denser; merchants and traders attended fairs and travelled extensively to transact their business; peddlars and chapmen journeyed to sell their wares; drovers delivered livestock; specialist craftsmen were drawn to distant places of work; purveyors made journeys to take up military supplies or goods for the royal household; migrant workers, apprentices, scholars, soldiers, paupers, vagrants, entertainers, lawyers and judges all undertook both short and lengthier journeys. Prisoners were conducted to county gaols. Officers of the Church, secular officers of parishes, towns and cities, messengers, post-boys, and servants conveying letters would also have been encountered on the highway.[14]

To support this increasingly mobile society, a whole infrastructure of facilities and services developed including signposts and inns for overnight stops. Inns were distinct from alehouses and taverns in providing accommodation for travellers. Inns had developed from the late Middle Ages but the increase in physical mobility brought about by the horse saw them thrive. In 1577, extrapolating from a large survey across 26 counties it has been estimated that there were around 3,600 inns in England out of a total of 20–24,000 alehouses, inns and taverns. By 1630, total drinking outlets had increased to 30,000, 40,000 by the 1680s and as many as 60,000 by 1700. Assuming the proportion that were inns was broadly constant, there were an estimated 6,000 to 7,000 inns in England by the beginning of the eighteenth century (although the proportion of drinking outlets that were inns may have risen). By the early eighteenth century, London's major inns numbered around 200.[15] Inns were located along all major roads, and clusters developed every 10–15 miles. The spatial clustering of inns meant that economic activities developed endogenously as well as exogenously through the passing trade on roads. Innkeepers were key in providing stagecoach services and also private hire of horse and carriages from the 1740s. Through the stabling of horses and provision of accommodation, the development of inns became a significant driving force in the small-town settlement structure along internal trade routes during the seventeenth and eighteenth centuries. Not only did the inn accelerate the potential for the expansion of the stage-coach system, but by the middle of the eighteenth century it effectively controlled that development as most coach-service providers were owned by inn-keepers. Inns therefore represented "the core entrepreneurial force" in passenger carriage.[16]

As personal mobility grew, so did demand for printed maps and schedules to help people make their way. Growing numbers of books were produced on horses and horsemanship. A network of carrier services was in place in the sixteenth and seventeenth centuries and operated on local, regional and national routes. Regular links between more distant places were weekly or fortnightly, but more frequent within counties and regions. Private carriers also operated on an ad hoc basis transporting goods and people as required. These networks helped establish an early system of horse-dependent interconnectedness, and the to-ings and fro-ings brought news and gossip as well as people and things. A surviving brokerage book from Southampton showed laden carts heading from the port for Worcester, Coventry, Manchester and Kendal, as well as Winchester, London and Oxford, and records from the University of Oxford show passenger carriages in the 1670s making regular trips to Lincoln, Somerset and Winchester.[17]

London had even more regular services. A guide produced by John Taylor in 1637 listed over 200 carriers and waggoners and the terminals they used. From *The George* near Holborn Bridge, travellers left for Buckingham, Brackley and Banbury on Wednesdays, Thursdays and Fridays. The Colchester carrier would depart from the *Cross Keys* in Gracious Street on Fridays. There were weekly carriers to and from Chard in Somerset, twice weekly to and from Cambridge, Chester and Coventry. By 1681, 645 carriers could be identified operating from London. Early stage-coach services, where horses were regularly switched in order to speed travel, date back to at least the mid-seventeenth century. First references to such services from London date from 1653 (York), 1654 (Cambridge, Lincoln, Worcester), 1655 (Winchester, Exeter), 1657 (Bristol, Dover, Norwich, Portsmouth) and 1658 (Edinburgh, Plymouth, Wakefield).[18]

Horse travel to convey correspondence was via accredited messengers and couriers. In the early sixteenth century a royal delivery system of standing posts was developed which helped regularise and speed up delivery of letters. This was designed for official Crown and Government use, but after 1635 it was opened up to the delivery of private letters. The standing posts were officials appointed at points along main routes to convey the post. In 1556, the main route from London to Berwick upon Tweed involved 25 posts including those based at Huntington, Grantham, Darlington, Newcastle and Alnwick.[19] By the late 1590s, there were 79 standing posts along 1,200 miles of road around the country.[20] They began to be used to offer horse-hire services to long-distance travellers but also served as a major stimulus to letter-writing. Private letters were carried by horse-back at rates of between two and three pence per mile. The network was vital in integrating domestic national markets and was crucial in achieving what Brayshay describes as a "joined-up realm".[21] The system was extended to Ireland and Scotland and to the English colonies in North America.

By the end of the seventeenth century, the network was also being used to distribute publications such as the *London Gazette*, which would reach the farthest corners of England within days of publication and "symbolises the new mobilities upon which the modern age was built".[22] News better travelled. Events reported in Livorno, Italy on 2nd June 1699, Warsaw on 6th, Vienna on the 7th, Berlin on the 13th, Brussels on the 17th and The Hague on the 20th were published in London on the 22nd and could reach Berwick or Plymouth a matter of days later. Horses underpinned an increasingly sophisticated level of national interconnectedness and the period between 1500 and 1700 saw a step change in the degree of interlinkages and in the volume of travel and communication.

The Horses at Work in the City

As we saw in Chapter 3, horses had played a vital role in industrial production, and in its transformation during the early part of the industrial revolution prior to the introduction of the steam-engine in the 1780s. Thomas Almeroth-Williams' study of animals in Georgian London, *City of Beasts*, charts the changing role and significance of different forms of work by horses in the burgeoning industrial economy of what was to become the first 'world city'.[23] His work counters the conventional innovation-centric approach to the history of the industrial revolution by highlighting the crucial process of labour becoming more productive. The ways humans worked with horses were critical in enhancing productivity and so horses can be placed at the heart of the origins of London's industrial development and of the industrial revolution more generally.

Crucial to the productive relationship between humans and horses in industrial production was the horse mill, which comprised a vertical rotating shaft, with horizontal beams to which a horse was harnessed and would walk around in circles driving the shaft. Mill horses were often older horses nearing the end of their working lives or rejects from the hackney and stagecoach trades. Through their work mill horses were powerful agents driving the development of London in the early industrial revolution. From newspapers, court and insurance documents, Almeroth-Williams identified more than fifty horse-powered mills in London across twelve different trades during the period 1740 to 1815, but these would represent only a fraction of the total. They were concentrated around the edge of the old city in areas like Old Street, Clerkenwell, Shoreditch, Smithfield, Southwark and Bermondsey. They were most commonly used to process raw materials and featured in the paint and dye industries, tobacco, brewing, glass-making, pottery and brick-making. Mills could also be found in the textile industry, paper presses, and sheet metal manufacture. They also played a significant part in the leather industry, which required relatively little plant and machinery but had hitherto relied on human strength and manual skills. As tanners sought to expand production and reduce labour costs, horse mills became vital. Bark from oak trees was used to extract tannin to convert raw hides into leather. A horse-powered mill suited these purposes and by the 1790s, insurance records show most of Southwark's tanners had mill horses and stables on their premises. This was a significant part of the London economy and by the 1820s, the leather industry accounted for about 1,000 firms and 7 per cent of London's manufacturing businesses.[24] Even with the advent of the steam engine, horses were initially retained in the tanning industry because they were so effective.

Horse mills were used in the production of paint and dyes and London's colourmen, or house paint manufacturers, were in the vanguard of the development of London as a wealthy metropolis. The period 1775 to 1785 saw a rapid expansion in affluent residential London with hundreds of new terraced houses built on the Portman, Portland and Bedford estates in Bloomsbury and the West End. Extensive further growth of the city's population and urban footprint took place over subsequent decades. Brightly painted home interiors became the fashion, paint became a 'hot commodity', and the industry boomed. By the 1820s, there were over 100 paint retailers in London and at least 42 colour-makers producing paint. The horse mill was a pivotal source of power in the shift to the mass production of paint.[25] Almeroth-Williams shows how the growth in paint production would not have been possible without horse-power and how companies would actively market themselves by promoting the role of horses in their production processes via trade cards and advertisements.

Another major industry where horsepower played a central role was brewing, the largest part of the food and drink manufacturing sector. Economic and population growth in London meant soaring demand for beer and the industry developed considerably during the eighteenth century. Between 1720 and 1799, there were between 140 and 180 firms in brewing and the industry became increasingly concentrated, with the twelve largest firms accounting for 42 per cent of production in 1748 and 85 per cent by 1830. These were massive companies and among Britain's biggest industrial operations at the time. Although the growth of the brewing industry is often associated with the growth of coal and steam power, the first steam engine was not used in the industry until 1784 and leading brewers were still using horses in the early nineteenth century. Horse mills were used for grinding malt, pumping water and drawing liquid wort, initially as separate processes. During the first half of the eighteenth century, breweries had developed mills which could pump and grind simultaneously. With this change, the "circular plodding of horses" increased in efficiency, enabling a surge in production and profits.[26]

Alongside the process of industrial production, horses played a central and critical role in the movement of people and goods. Almeroth-Williams describes draught horses as "at the very heart of the metropolitan economy" in Georgian London.[27] There were significant improvements in breeds over this period, particularly through cross-breeding with imported stock, and by the 1750s a handful of elite breeds dominated. A common draught horse was the Midland Black, originally from Leicestershire but with blood from Flemish imported stock. Strength and activity were more important than height or weight and selective breeding gradually shortened the Midland Black's chest, neck, back and legs but fostered a short, thick carcass and thick legs. Alongside the relatively elite Midland Blacks were a range of lesser draught horses used by traders and shopkeepers.

Haulage work for London's draught horses swelled massively over the eighteenth century, with the number of wagons in the city trebling between 1680 and 1830, by which time about a thousand were crossing the capital each week, bringing in food supplies, transporting raw materials and carting manufactured goods. The wagon firms, and the horses, were often based in provincial towns, although many horses were stabled within the city. Two-wheeled carts were allowed to carry a load of up to a one-ton limit, raised to 1.25 tons in 1757, but four-wheeled wagons could carry heavier loads. Carts were privately owned by tradesmen or were available for hire in the city from licensed carmen. There was considerable work from the construction industry hauling timber, bricks and stonework, slate, gravel and lime and horsepower fuelled a great construction boom in London. Timber was shipped in from elsewhere in the UK, as well as from Scandinavia and America, and had to be hauled by horse from the Thameside wharves to the building sites all over the city. Bricks were moved shorter distances as brick production was located closer to construction sites, but draught horses would be employed to ensure a steady supply. The stonework used in the building and sprucing up of the West End was hauled by horse, and there was considerable material required for the maintenance and improvement of London's roads and pavements, not least because of the wear and tear generated by the growth in heavy horse-drawn traffic.

A vast equine industry developed to manage waste from the city. Cesspools had to be emptied and soot swept from chimneys had to be disposed of. Nightmen provided specialist services in the transport and removal of "mountainous quantities of human faeces, animal dung, food scraps, cinders, rubble and industrial waste".[28] The trade complemented the importing into the city of gravel, clay and sand as waste was hauled out to rubbish tips and wagons returned laden with building materials. Specialist waste removal could generate a decent living and some nightmen hauliers built up significant fleets of carts and wagons. Small businesses could hire errand carts for specific haulage needs and luxury consumer goods could be delivered in this way to wealthy individual customers. Court records reveal errand carts being used to transport tobacco from Bishopgate to Woolwich, and stolen indigo to Paradise Row. Probably the most significant cargos in the eighteenth and nineteenth centuries, however, were coal and beer.

Coal power fuelled the industrial revolution, but much of its distribution was serviced by horses.[29] In 1640, 150,000 tons of coal was being brought by ship to London, but this had grown 8-fold by 1800 to 1.2 million tons, the equivalent of 400,000 wagonloads. The coal had to be hauled by horse-drawn wagons from Thames-side wharves uphill into the city. Independent coal merchants built up fleets of wagons and large stables. One Upper Thames Street merchant in the 1770s had 24 horses, nine carts and one wagon. Another in the 1840s had 36 horses, six wagons and four carts. By 1790, it is estimated that London's 203 coal merchants would have employed of the order of 2,000 horses. Draught horses were worked hard hauling coal and not only supplied domestic homes but also the industrial furnaces used in brewing and manufacturing industries such as soap-making, sugar-refining and glass-making. By the late nineteenth century, an average draught horse working in coal haulage might shift 30 tons of coal a day and work an 80-hour, six-day week. At the other end of the supply chain, in the coalfields of the north, Galloway pit ponies would drag sledges of coal through and out of the mines. Horse-drawn coal fuelled the development of cities and the modern life within them. The development of the railways drove up London's demand for coal, from 1.25 million tons in 1810 to six million in 1860s and eight million in the 1890s. To service this growth, the number of horses hauling coal in London rose to 5,000 by the 1840s and 9,000 by the 1890s.[30]

Dray horses were large, powerful draft horses used to pull the heaviest loads, usually on strong four-wheeled wagons. They were also prominent in the eighteenth and nineteenth century city and in London are estimated in 1800 to have accounted for over 1,000 animals among 40 breweries. Dray horses often featured in the paintings of the time and were an iconic image of the industrial revolution. "For Georgian Londoners, nothing conveyed the nation's industrial progress and prosperity more compellingly than the city's dray horses".[31] Midland Blacks were popular and would be expected to haul wagons with 1.5 tons of barrelled beer for twelve hours a day or more. Dray horses were widely admired by visitors to Britain, not just for their strength and physical appearance, but also for their calm temperament in an increasingly noisy and busy urban environment. They were renowned for their intelligence and spatial awareness and had good memories for patterns, routes and work routines. The work by dray horse increased significantly over the Georgian period particularly as breweries took over pub leases and began delivering regularly to publicans across the city. It was not unknown for dray horses to work sixteen hours in harness. Improved feed and stabling meant productivity improved over time. In the 1770s, a dray horse could haul 1,433 barrels a year, but this increased to over 2,600 by the 1820s.[32]

The Horse and Leisure in the City

In addition to the world of work, horses played an instrumental role in leisure and recreation in cities and horse-society became the height of fashion in eighteenth-century London and the focus of a British equestrian obsession. The culture of the horse became a central feature of fashion and sociability among the upper classes and aspirant merchant classes and a range of practices and institutions developed around the horse that endured for more than a century and whose imprint can still be observed in the contemporary city. In addition, Georgian Londoners' equestrian leisure, what Almeroth-Williams calls 'horsing around', helped develop new types of social and spatial relationships between urban and rural areas that began to reconfigure and reposition cities within their hinterlands. Horse culture led to the construction of large spectator venues in the city such as the circus and the riding ring and spawned the popular pastime of a ride out to the countryside.

London's upper classes split their time between the town and the country, with indoor sociability dominating the former and riding and hunting the latter. However, horse-riding did feature as a pastime in the city and in the second half of the eighteenth century it has been argued that equestrian culture "became one of London's most successful recreational departments".[33] The wealthy stabled increasing numbers of horses and carriages in their homes and the less wealthy were able to hire a saddled horse or a one-horse chaise by the day. Some servants who looked after horses were also able to exercise them, so some horse and horse-drawn traffic involved people simply out for a ride. As the city's horse stock increased, so the commercialisation of equestrian recreation expanded too and brought with it the establishment of all sorts of venues to attract recreational visitors.

Hyde Park in London's West End became the focal point for elite and ostentatious equestrian leisure. A wide ring was railed off for horse-drawn carriages to ride around which attracted those wealthy enough to own coaches and carriages. Riding around showing off horses, vehicles and tackle became the height of fashion. Activity centred on what was known as Rotten Row. Over time, ever larger crowds were attracted to the park, especially on Sunday afternoons, to either ride or watch. By the early nineteenth century, the attraction had become so popular that a grand scheme was devised to equestrianise the park and integrate the various drives and ride-ways into a complete formal circuit. Parading around the park was a key site in which people could show off their riding skills and abilities.

Being able to ride well, and ride with refinement, became a sign of social standing and this stimulated the development of public riding schools in London. These promoted the continental art of managed riding, akin to modern day dressage, and became an important rite of passage for the young of the elite. Between 1760 and 1835, at least 26 new public riding schools were established in London, and they endured for a few decades. There were still twelve in the 1820s and they did not die away until the 1850s.[34] In the late 1780s, the number of schools trebled, and several were located around Hyde Park Corner and served the wealthy new residents of the expanding West End. Some also developed around the City and Lambeth to provide services for the growing mercantile classes. Students generally learned the discipline of managed riding, with an emphasis on poise, grace and skill, but also safety. The Earl of Chesterfield wrote to his nineteen-year-old son in 1751 urging him not to "neglect your exercises of riding, fencing, and dancing … for they all concur to *dégourdir* [smooth rough edges] … To ride well, is … a proper and graceful accomplishment for a gentleman".[35] Riding schools were fashionable and influential and played a catalytic role in the development of equestrian recreational practices.

In the latter half of the eighteenth century, enjoying a ride in the suburbs and countryside, or 'riding out', became an increasingly popular leisure pastime which had significant implications for spatial practices such as the rise of domestic tourism and of commuting. Riding out required owning or hiring horses, vehicles and tackle, but was less expensive than riding schools or the ostentations of Hyde Park and so more widely popular among the middling classes. It became a means of identity formation among more affluent middle-class men, although was ridiculed by satirists who caricatured what they saw as desperately aspirational city types taking on the pretensions of the rural landed classes. This draw of the countryside among aspirational middle-class London gentlemen was an early forerunner of the development of idealised and romantic notions of the countryside as a civilised retreat from urban life.[36] Elegant villas began to be built in the countryside around London from which merchants and financiers could commute into the city by horse.

Horse-racing was also a popular pastime. The sport has a long history dating back at least to Roman times when horse-drawn chariots were raced as a spectator sport.[37] In Britain, racecourses began to be developed in the sixteenth century, the first being at Chester. Racecourses were usually developed by towns and cities on land owned by corporations, and courses were developed at Richmond, York, Boroughbridge, Doncaster and Carlisle all during the sixteenth century. Racing at Newmarket developed after King James I stumbled across the town in February 1604, developed it as a holiday retreat and it became a key centre for royal stables and breeding. Racing at Ascot, Epsom, Lincoln and Enfield also developed as a result of royal interest and patronage.[38] During the eighteenth century, horse racing boomed to become the most significant leisure event of the summer season and the first 'proto-modern' sport in Britain. By the early nineteenth century, horseracing was deemed to be *the* sport of Britain's people, its national sport.[39] Its popularity was rooted in the experiences and narratives it offered about successful horses, trainers and breeding and the opportunities it provided for pleasure and sociability in a tacitly controlled cross-class environment. It attracted a more socially diverse pattern of participation than other sports of the day and this was part of its appeal. The development of the sport was linked to the growing commercialisation of leisure in general and especially among the growing middle classes. Changes in transport, the growth of newspapers and the growing individualistic spirit of the age all helped underpin racing's growing popularity. Betting helped fuel the horseracing economy and a set of institutions and rules were developed to codify and oversee the operation of the sport.[40]

The increasing popularity of riding out among Londoners coincided with the growth of horse-racing and by the second half of the eighteenth-century thousands of horse riders from the city would ride out to surrounding racecourses. Racing had expanded dramatically between the 1680s and the 1730s. In 1700 the only racecourses within 32 km of London were Barnet, Croydon and Epsom, but by 1738 there were almost 20, several of which were within 8 km of Charing Cross (including Kentish Town, Belsize, Hampstead and Highgate). Prior to 1740, most racecourses were within walking distance of the city, but by the 1760s they were located further afield, and so Londoners would ride out to the races. Large crowds would gather at racing events, with an estimated 30,000 attending racing at Barnet in a single week in 1771.[41] Horse and vehicle rentals boomed during the racing season and the roads would be lined with carriages all the way from London.

Deeper into the countryside Londoners could hunt with hounds and had been doing so with hunts in the Home Counties since at least the early eighteenth century. Fox-hunting became fashionable during the eighteenth century among those wealthy Londoners who enjoyed the pretensions of lording it on horses with the rural gentry as well as the thrill of the chase. It was an expensive pastime, but London money helped sustain a large number of hunting packs around the edge of the city including Bermondsey, Putney and Leatherhead. As the city grew larger so hunting was forced further afield. The longer distances to travel to a hunt made riding out for a day's hunting less viable as horses tired, so some London hunters would hire local hunting horses, and others bought properties in the countryside close to hunts.

Hunting, racing, parading around Hyde Park, gathering at riding schools, or just riding out, was all means by which the institutions of horse culture enabled Londoners to come together and socialise. Riding out could be combined with trips to taverns by men and was also a means by which men and women might socialise together, although in polite metropolitan society these encounters would be carefully managed. Both wealthy men and women found riding out a thrilling and exciting part of social life, largely through the social interactions between people, but also because of the joys and freedoms that come from riding a horse. Favoured destinations for Londoners riding out included Hampstead, Muswell Hill, Tottenham, Battersea and Clapham, all scenic villages amidst rural terrain that riders could enjoy from an elevated view atop a horse. "Horses liberated Londoners by expediting access to private space but also because they greatly enhanced the experience of enjoying the countryside".[42]

London was central to England's horse culture in the eighteenth century and new equine socio-spatial practices and meanings were developed in the city that spread far more widely. London increasingly dominated British society and culture, but equestrian society was even more geographically concentrated in London than was wider society. By the mid-eighteenth century, London was home to almost half of England's 9,000 four-wheeled private carriages and almost 3,000 further two-wheelers. By the early nineteenth century there would have been well over 30,000 horses in the city being ridden principally for pleasure.[43]

Horses, Urban Form and Environment

This far we have tended to focus on the growth in the use of the horse in industry and for transport as part of the progress in economic development, greater interconnectedness and sociability. However, the increasing use of the horse also brought social costs and inconveniences to urban life. The expansion of manufacturing and the growth of trade in the early modern period led to a growth in wheeled traffic through town streets that had developed in the medieval period and congestion became quite a problem. Emily Cockayne produced a study of the 'hubbub' of the period and the responses of disgust and frustration it provoked.[44] Where hackney carriages clustered this caused difficulties for merchants and other travellers. Horse-drawn vehicles need a wide turning circle, but this was often problematic in densely packed urban areas. Conflicts also arose between horse-drawn traffic and pedestrians, not helped by the fact that streets were not well demarcated. The growth of traffic became such a concern that some livery corporations in 1671 called for stagecoach numbers to be regulated in London. By 1720, carts and coaches were considered to be the main cause of death and injury to people on London's streets. Children were particularly at risk because they were less visible. Cockayne found that of 70 cases of cart, coach, dray and wagon accidents that caused fatal injuries between 1679 and 1770, over half the victims were children, with carts particularly lethal. Details are contained in Old Bailey reports:

> In 1691 Charles Collins accidentally killed a young girl called Sarah Smallnick by driving his brick cart 'over her Short Ribs' .... A little boy 'was peeping at the Corner of the Post' shortly before he was trapped between the post and a cartwheel in 1696. He died of his injuries. One child was killed as he knelt down to 'take some Mulberry Leaves off the Street when the Dray went over his Head, and broke his Skull'. A toddler stooping to pick up gravel stones from the road in 1727 was felled by a cart. Another child was hit while playing football in the street. In one particularly tragic case a child was playing with straw on the road, and 'being cover'd with the same' was not spotted by a cart driver until it was too late.[45]

Often, vehicle drivers blamed the skittishness of horses and were sometimes acquitted, although there was some public concern about traffic not being driven carefully enough and much complaining about the rough and antisocial attitude of many hackney carriage drivers. The busiest parts of towns tended to have the narrowest streets which compounded congestion problems and undermined safety. Marketplaces were usually located in these most constrained parts of towns and cities. In the late seventeenth century, residents living near some of London's markets petitioned Parliament to more tightly restrict access by country folk bringing in their wares and causing congestion.[46] Efforts were made to restrict and regulate street furniture and stalls on streets, including in Oxford, Bath and Manchester, and some parts of markets were moved elsewhere to free up space and ease the flow of horse-drawn traffic. Bridges and town and city gates were acute pinch points for horse-drawn traffic. Sometimes regulatory efforts were made to improve the flow of traffic by limiting the number of vehicles crossing at any one time. Often, new, larger bridges had to be constructed and old medieval gates dismantled and whole new roads were built to cater for horse traffic. In Bath, for example, three of the city's medieval gates were demolished in the mid-eighteenth century to improve urban planning and traffic flow.[47] In London in the early 1760s, the gates at Ludgate, Cripplegate, Aldgate, Aldersgate, Bishopgate and Moorgate were all removed, all to improve the smooth flow of horse-drawn carts.[48]

Another source of nuisance for the urban environment in the horse-drawn city was dung, and horses produced copious amounts. Dung was a valuable resource to use as manure and so was generally collected, but amounts were so voluminous that horse dung inevitably also mixed with the general dirt and detritus of the street. From the early seventeenth century some towns and cities employed municipal street cleaners. Scavengers would collect waste by cart that might be spread on common ground or taken to waste dumps at the edge of the city. Sometimes local residents would club together and employ staff to keep streets and public spaces clean. Waste was piled into dunghills –

> a stinking morass of human and animal waste, rotten timber, friable plaster, rubble, carcasses, cinders and ashes, broken glass and crockery, clay pipes, spent bedding, feathers, straw, weeds, eel skins, fish heads, peelings, husks, stalks and cores, in various states of decay.[49]

They were used as public toilets at night and chamber pots were often dumped onto them. Unsurprisingly, they gave off the most offensive stench, especially in the summer months. As dunghills and muckheaps were cleared away, usually hauled by horses, material was dumped around the edge of towns and cities. Waves of expansion then meant that the urban footprint grew over its own filth.

We saw earlier how horses played a key role in the construction industry that built the towns and cities of the fifteenth to nineteenth centuries by hauling material to and from construction sites. As the urban horse population grew in the eighteenth and nineteenth centuries, so cities became moulded around the horse in myriad ways. Riding schools, grand arenas for circuses, and markets for buying and selling horses are some examples of key equestrian developments. Pickfords' large stable complex for its working horses on the site of what is now Camden Market provides another. The many trades and businesses that supported the horse economy would also have been prominent features of urban life, with blacksmiths and farriers, saddle-makers, coach-builders, wheelwrights and wainwrights spread throughout the city. Horses also crucially influenced the built environment through the systematic development of stabling within the new residential areas built within the growing cities and through the widening and enhancing of thoroughfares to accommodate increasing horse-born and horse-drawn traffic.

To house the private horses of more affluent households, thousands of stables and coach houses were built in the form of mews complexes. London's first mews were developed in Covent Garden in the 1630s, but most were built after 1720. There were 29 examples of mews complexes by the 1740s and 117 by the early nineteenth century.[50] They were a key feature of the major development of the estates that now comprise London's West End, from Bloomsbury through Marylebone and Mayfair, Bayswater and Notting Hill, and Belgravia and Kensington. Henry Mayhew in 1851 wrote:

> The mews of London constitute a world of their own. They are tenanted by one class – coachmen and grooms, with their wives and families – men who are devoted to one pursuit, the care of horses and carriages; who live and associate one among the other; whose talk is of horses (with something about masters and mistresses) as if to ride or to drive were the great ends of human existence.[51]

The mews offered a system for housing people and horses in an integrated complex that allowed main residences to front onto the street, with stables, coach houses and hay lofts accessed by a separate alley. Mews were discretely screened by arches set into principal street facades which both enhanced the streetscape architecture and hid the mews. They were able to be developed on an extensive scale because large swathes of west London were developed in large blocks and at speed. The Bedford and Grosvenor estates were developed first with a similar pattern of planning and building on long leaseholds. The model was followed for the Holland and Ladbroke estates and then much more widely. The overall ratio of mews units to houses was 1: 1.4 in the Grosvenor estate and in the best residential parts of London there were probably as many stables as houses.[52]

The development of the West End estates was designed in an ordered grid to accommodate horse drawn carriages, with the mews as the means of keeping horses and carriages close by but off the streets when not being used. Larger mews units were equine palaces that could accommodate as many as 24 horses.[53] Housing for horses became an important factor in the attractiveness of the new streets and estates of the expanding West End. The need to accommodate horses featured prominently in newspaper adverts for wanted properties. By the early nineteenth century, new large-scale developments in the West End were close to the point where every house had its own stable and coach house.

The horse was utterly pervasive in creating the sounds, sights and smells of the streets. Architect H.B. Creswell, who was born in 1869, wrote in the *Architectural Review* in 1958 about his childhood in the 1870s and 1880s.

> The whole of London's crowded wheeled traffic – which in parts of the city was at times dense beyond movement – was dependent on the horse: lorry, wagon, bus, hansom and 'growler', and coaches and carriages and private vehicles of all kinds … Meridith refers to the 'anticipatory stench of its cab-stands' on railway approach to London with gay excitement – was of stables … whose middens kept the cast-iron filigree gas chandeliers that glorified the reception rooms of middle class homes … encrusted with dead flies and, in late summer, veiled with jiving clouds of them. A more assertive 'mark of ze 'orse' was the mud that, despite the activities of a numerous corps of red-jacketed boys who dodged among the wheels and hooves with pan and brush in service to iron bins at the pavement edge … flooded the street with churnings of 'pea soup' that at times collected in pools over-brimming the kerbs and … covered the road surface with axle grease or bran laden dust to the distraction of the wayfarer. In the first case, the swift-moving hansom or gig would fling sheets of such soup – where not intercepted by trousers or skirts – completely across the pavement, so that the frontages of the Strand, throughout its length had an eighteen-inch plinth of mud-parge thus imposed upon it. The pea-soup condition was met by wheeled 'mud carts' each attended by two ladlers clothed as for Icelandic seas in thigh boots, oilskins, collared to the chin, and sou'westers sealing the back of the neck. Splash Ho! The foot passenger now gets the mud in his eye! … Hence arose London's shoe-blacks, registered and red-coated, at the kerbside doing a mighty job with blackening and spittle for 2d; and hence also London's crossing-sweeper of the fashionable quarters. And after the mud the noise, which, again endowed by the horse, surged like a mighty heart-beat in the central districts of London's life. It was a thing beyond all imaginings … the hammering of a multitude of iron-shod hairy heels upon [the streets], the deafening, side-drum tattoo of tyred wheels jarring … like sticks dragging along a fence; the creaking and groaning and chirping and rattling of vehicles, light and heavy, thus maltreated; the jangling of chain harnesses and the clanging or jingling of every other conceivable thing else, augmented by the shrieking and bellowing called from those of God's creatures who desired to impart information or proffer a request vocally – raised a din that, as has been said, is beyond conception. It was not any such paltry thing as noise. It was an immensity of sound ….[54]

The expansion of horse-drawn traffic in the eighteenth and nineteenth centuries also necessitated further rationalisation of the networks of streets in London and elsewhere. An integrated system of carrying services had been developed by the mid-eighteenth century covering most large urban centres. In 1767, Birmingham had 160 services a week, provided by 54 different firms, linking to King's Lynn, York, Newcastle, Lancaster, Manchester, Shrewsbury, Bristol and Southampton, as well as London. Norwich had 59 services a week in 1729. Bristol had 148 by 1755.[55] By the 1760s, larger wagon vehicles were being increasingly used, pulled by eight horses and sometimes loaded with as much as six tons of goods.[56] The massive growth in horse-drawn road traffic, and the greater size of wagons and loads, meant towns and cities had to be reshaped around the horse. Streets were widened and straightened, and their surfaces had to be managed.

'De-horsification': The Disappearance of the Horse from Urban Britain

As we saw in Chapter 3, the industrial revolution brought the onset of the decline of the horse from urban areas, although this was not a simple and straight-forward decline. The development of the railways heightened the need for horses to ferry goods and people to and from railway stations and so the first few decades of the railway age saw continued growth in horse numbers in the UK. In the twentieth century, horses began to disappear from urban streets quite rapidly from the 1920s onwards as they were replaced by motor vehicles. The decline in horse numbers in rural areas lagged behind. This pattern was replicated in other advanced nations.[57] In 1881, a New York Times editorial had emphasised the indispensability of the horse in the functioning of the city. "Deprived of their human servitors, the horses would quickly perish; deprived of their equine servitors, the human population in cities … would soon be in straits of distress".[58] Yet things changed very quickly in the urban sphere. Not only did the decline in horse numbers lead to a major material change in how towns and cities functioned and in the urban landscape, but it also changed the ways that horses were perceived and represented. Strong and smart dray horses pulling brewery wagons had been the height of industrial modernity, but the petrol and diesel engine became the epitome of the new twentieth century modernity. There was such rapid change in technology and the urban economy that the economic adjustment through loss of livelihoods among the many who worked with horses never registered as a social or economic problem. Petrol stations, garages and motor mechanics took the place of stables, wheel- and wainwrights and farriers. Working horses soon became relegated first to rural life, and then to rural history, while seemingly being phased out of urban history altogether. The remaining relics of urban equine life were the statues of mounted generals and the occasional teams of police horses required to police large crowds and, in London at least, the pageantry of the royal household cavalry and their proud trotting up and down the Mall, with tassels swinging atop shiny helmets. Cities that in the eighteenth and nineteenth century had become increasingly 'horse-shaped' now began to evolve around the needs of the motor car and diesel lorry.

However, the horse left its mark on our cities, and not only in the form of statues holding up military heroes of ages gone by. Even in the 1980s, 600 mews complexes remained in London still contributing to the city's urban form and civilised atmosphere. With the great bijou gentrification of the late twentieth and early twenty-first century, these cobbled enclaves became highly sought-after for housing the affluent. Both central and out of the way, the mews became a model for new residential development, just without the horses.

Notes

1 de Vries, J. (1984) *European Urbanisation 1500–1800*. London: Routledge.
2 McShane, C. and Tarr, J. (2007) *The Horse in the City: Living Machines in the Nineteenth Century*. Baltimore: Johns Hopkins Press; see also Greene, A. (2008) *Horses at Work: Harnessing Power in Industrial America*. Cambridge, MA: Harvard University Press.
3 Greene (2008).
4 Brayshay, M. (2014) *Land Travel and Communications in Tudor and Stuart England: Achieving a Joined-Up Realm*. Liverpool: Liverpool University Press, p.19.
5 Wrigley, E. (2006) The transition to an advanced organic economy: half a millennium of English agriculture, *Economic History Review* 59, 435–80.
6 Harvey, E. (2010) Pavage grants and urban street paving in medieval England, 1249–1462, *Journal of Transport History* 31, 151–63.
7 Glennie, P. and Thrift, N. (2009) *Shaping the Day: A History of Timekeeping in England and Eales, 1300–1800*. Oxford: Oxford University Press.
8 Brayshay (2014), p.124.
9 Brayshay (2014), p.91 & p.107.
10 quoted in Brayshay (2014), p.111.
11 Mortimer, I. (2023) *Medieval Horizons: Why the Middle Ages Matter*. London: Bodley Head pp.140–41.
12 Brayshay (2014), p.200.
13 Brayshay, M. *et al.* (1998) Knowledge, nationhood and governance: the speed of the Royal post in early-modern England, *Journal of Historical Geography* 24, 265–88.
14 Brayshay (2014), pp.349–50.
15 Chartres, J. (2002) The eighteenth century English inn: a transient 'golden age'?, pp.205–26 in Kumin, B. and Tlusty, B. (ed.) *The World of the Tavern: Public Houses in Early Modern Europe*. Basingstoke: Ashgate, pp.207–08.
16 Chartres (2002), p.219.
17 Brayshay (2014), p.132 & p.140.
18 Brayshay (2014), p.141 & p.145.
19 Brayshay (2014), p.276.
20 Brayshay, M. (1991) Royal post-horse routes in England and Wales: the evolution of the network in the later-sixteenth and early-seventeenth century, *Journal of Historical Geography* 17, 373–89.
21 Brayshay (2014).
22 Brayshay (2014), p.347.
23 Almeroth-Williams, T. (2019) *City of Beasts: How Animals Shaped Georgian London*. Manchester: Manchester University Press.
24 Almeroth-Williams (2019), pp.20–23.
25 Almeroth-Williams (2019), pp.25–29.
26 Almeroth-Williams (2019), pp.35–39; see also Almeroth-Williams, T. (2013) The brewery horse and the importance of equine power in Hanoverian London, *Urban History* 40, 416–41.
27 Almeroth-Williams (2019), p.41.
28 Almeroth-Williams (2019), p.49.
29 Paxman, J. (2021) *Black Gold: The History of How Coal Made Britain*. London: William Collins
30 Almeroth-Williams (2019), pp.51–3.
31 quoted in Almeroth-Williams (2019), pp.54–6.

32 Almeroth-Williams (2019), p.61.
33 Almeroth-Williams (2019), p.160.
34 Almeroth-Williams (2019), p.163.
35 quoted in Almeroth-Williams (2019), p.165.
36 Lowe, P. *et al.* (1995) A civilised retreat? Anti-urbanism, rurality and the making of an Anglo-centric culture, pp.63–82 in P. Healey *et al.* (eds.) *Managing Cities: The New Urban Context*. Chichester: John Wiley and Sons.
37 Hyland, A. (1990) *Equus: The Horse in the Roman World*. London: Batsford.
38 Tomlinson, R. (1986) A geography of flat-racing in Great Britain, *Geography* 71, 228–39.
39 Huggins, M. (2018) *Horse Racing and British Society in the Long Eighteenth Century*. Woodbridge: Boydell Press, p.278.
40 Huggins (2018).
41 Almeroth-Williams (2019), p.170.
42 Almeroth-Williams (2019), p.182.
43 Almeroth-Williams (2019), pp.155–6.
44 Cockayne, E. (2007) *Hubbub: Filth, Noise and Stench in England 1600–1770*. London: Yale University Press.
45 Cockayne (2007), pp.170–71.
46 Cockayne (2007), p.175.
47 Cockayne (2007), p.178.
48 Almeroth-Williams (2019), p.40.
49 Cockayne (2007), p.188.
50 Almeroth-Williams (2019), p.140.
51 Quoted in Rosen, B. and Zuckermann, W. (1982) *The Mews of London: A Guide to the Hidden Byways of London's Past*. Exeter: Webb and Bower, p.10.
52 Rosen and Zuckermann (1982), p.22.
53 Almeroth-Williams (2019), pp.140–41.
54 H. B. Creswell (1958) Seventy years back, *Architectural Review*, December 1958, quoted in Rosen and Zuckermann (1982), p.27.
55 Barker, T. and Gerhold, D. (1993) *The Rise and Rise of Road Transport, 1700–1900*. Basingstoke: Macmillan, p.27.
56 Barker and Gerhold (1993), p.40.
57 Geels, F. (2005) The dynamics of transitions in socio-technical systems: A multi-level analysis of the transition pathway from horse-drawn carriages to automobiles (1860–1930), *Technology Analysis & Strategic Management* 17, 445–76.
58 McShane and Tarr (2007), p.178.

5 Horses and the Country

Introduction

The horse has been an important agent in shaping the rural landscape over centuries, mostly as a working animal in agriculture and rural industries, but also in its role as an animal to ride for sport and pleasure. The legacy of equine work and play has left an imprint on the landscape that endures today. The story of the horse in British agriculture is book-ended with two transitions. The first is the replacement of the oxen by the horse in the eleventh to sixteenth centuries. This was a gradual and protracted process as horses were first introduced alongside oxen, then began to be more extensively used in some parts of the country, and eventually took over as the principal agricultural draught animal. The second transition took place in the twentieth century and is the replacement of the horse by the tractor. In historical terms, this twentieth-century transition away from the horse in agriculture was much more rapid than its gradual introduction centuries earlier but was a slower and more protracted process than the fall away in horse numbers used in passenger and freight haulage in the towns.[1] The agricultural workhorse gradually faded away between the First World War and the late 1950s, to become a feature of rural museums and a spectacle at country shows.

The growth in horse numbers over several centuries up to the early twentieth century necessitated the development of an extensive horse economy supporting the livelihoods of horse breeders and traders, blacksmiths and farriers, saddlers and so on. The growing requirement for horse feed meant that supporting horses on farms, as well as in their burgeoning numbers in cities, was a major influence on rural land use as fodder crops were required to feed a horse population that peaked at over 3 million. Alongside the world of agricultural and haulage work, horses were also part of rural leisure pursuits such as riding, racing and hunting, which have long been cited as important activities in maintaining the social fabric of some rural communities and which have also influenced the management of the rural landscape. From the 1980s, an expansionist agricultural policy was gradually replaced with greater emphasis on the countryside as a place of consumption which helped reposition the horse as a force for rural leisure and economic diversification. As a result, although horse numbers in rural areas initially

DOI: 10.4324/9781003454359-5

declined during the twentieth century they began to pick up again in its latter decades as equestrianism prospered as a sport and pastime.

This chapter examines how the relationship between horses and people in the countryside has evolved and the impact horses have made on rural geography. The chapter describes the role of horses in agriculture and its evolution over recent centuries. It then explains the evolving importance of hunting with horses and hounds in the social structure of rural Britain before examining the influence of horses in the management of the rural landscape. It concludes with an account of the changing relationship between humans and horses since the mechanisation of agriculture from the mid-twentieth century and the shift to a countryside more oriented to recreation and leisure.

Horses in Agriculture

Among the first steps in the development of agriculture as an alternative to hunting and gathering was the domestication of goats and sheep. Next came the domestication of wild cattle which were found across the Eurasian landmass. The ox was first domesticated in the Near East around 6,500 BCE and castration was being practised to produce docile and powerful oxen by Neolithic communities in central Europe by the fifth millennium BCE. There is evidence that oxen were pulling ploughs in western Europe by 3,600 BCE[2] and they were doing so in Britain before Roman times. Horses took time to replace oxen as agricultural draught animals. There is evidence of large stud farms in the Anglo-Saxon period in Suffolk, Essex and Staffordshire, although horses were still a luxury beast in the early medieval period. It is thought that the gradual replacement of oxen by horses began around the twelfth century, but it was not a simple or straightforward process and varied geographically.[3] The relative advantages of the horse over oxen depended upon soil type and terrain.

Records at the time of the Domesday Survey in the eleventh century suggest that workhorses in England represented around 5 per cent of all livestock in the early medieval period, equating to one horse for every 19 oxen.[4] By the fourteenth century, the average frequency of work horses in England had risen to around 27 per cent on the nobility's estate lands. Although oxen still numbered between two and three per horse, in some eastern counties the proportion of horses was higher.[5] Horses were used in teams mixed with oxen which helped increase ploughing speed while still retaining the strength of the oxen. These mixed teams were particularly common in south east England and East Anglia, but less so in the south west, west Midlands and the north. Horses were better suited to the south and east because they could be fed grain there. They were more versatile than oxen and could be used for hauling, harrowing and riding and as pack animals. They could be bought more cheaply too because they did not retain any value as a source of meat in the way that oxen did. They suited lighter soils or stony land and were more commonly introduced in areas where there was less pasture.

The growth in the number of horses in the early modern period was for their uses both in the country and in the town. This growth was not universally seen as a positive development because of the effects on demand for land and the displacement of capacity for human food production. As late as the end of the eighteenth century, Nathaniel Kent, a land valuer and estate agent, conducted an agricultural survey of Norfolk in which he expressed concern at the growing number of horses. He wrote:

> The more the number of horses can be lessened, the better, for all ranks of people. The consumption by horses, especially horses of pleasure and luxury, is astonishing; for though a horse in agriculture does not consume above three acres of the fruits of the earth in a year, a horse kept upon the road eats yearly in hay and oats, the full produce of five acres of land ... I am told, and I believe from good authority, that in the city of Norwich, not quite fifty years since, there were only twelve carriages of pleasure and luxury, and that there are now seventy-two, including post-chaises, and thirteen hackney coaches besides; and if we allow three horses to each carriage, upon an average, allowing for change, this will make a difference of two hundred and nineteen horses in the city of Norwich only. At that time, there was only one coach to London; now there are two mail coaches, and two heavy coaches; and as these cannot be allowed less than sixty horses to each mail coach, and fifty to each of the others, this makes an increase of one hundred and seventy horses more. There is also the coach to Lynn and another to Yarmouth which cannot take less than twenty horses more – here then is a difference upon a round calculation of four hundred and nine additional horses in what affects Norwich only; which at five acres to a horse, consume the additional produce of 2,045 acres.[6]

Despite these reservations, horses' numbers grew. The total number of horses in England and Wales grew by about 200,000 between 1770 and 1850.[7] Economic historian F.M.L. Thompson estimated that there were about 800,000 horses used on farms in 1811 (out of a total of 1.3 million in the whole country) and this number had increased to 1,511,000 (out of a total of almost 3.3 million) by 1911.[8] The number of horses in England and Wales recorded in the agricultural census peaked in around 1901, after which the number started to decline. The total population of horses in Britain was 3.3 million in 1901, although most of the increase during the nineteenth century was due to the growing need for horses in towns rather than in the countryside.[9] In agriculture, the key development was the growth in use of the tractor and its role in the mechanisation of draught power on farms. However, the replacement of the horse by the tractor in the twentieth century was more drawn out than might initially be expected. Tractors were first developed in the United States in the 1890s and were being mass produced by the 1920s. Yet, as late as 1950, tractors were still only providing a quarter of draught power in European agriculture, although they became the predominant force in Britain during the Second World War.[10]

In Britain, farming incomes were depressed during the 1920s and 1930s, and limited money to invest is one reason for the stilted early adoption of tractors, which were expensive to buy. In one of the first studies of the 'tractorisation' of British agriculture, the agricultural historian Ted Collins argued that the decline of the use of the agricultural horse in Britain between 1920 and 1939 was due to a more complex set of factors than just the adoption of the tractor, which initially displaced the steam engine rather than the horse.[11] Changes in statistical definitions complicate the issue, but it is undisputed that numbers of agricultural horses declined during the 1920s and had fallen by about a third by 1939. There was a geographical patterning to the process, and early declines in agricultural horse numbers were greater in the arable south and east and smallest in west and northwest England and in Wales.

Most horses used for agricultural work in the early twentieth century were heavy horses such as Shires and Shire crosses with other large breeds such as the Suffolk, Clydesdale and Cleveland. These horses would take at least 6 years from putting the mare to the stallion to having a fully mature and work-ready horse, although they would have a long working life, as a draught horse for up to 8 years and a further 6 to 10 years beyond that. These timelines meant that change could be slow. From the 1880s, most horses in Britain were employed outside agriculture, and by 1911, only 35 per cent of horses were occupied in farming.[12] The horse economies of town and country were interconnected and complementary. "One did the breeding and schooling, the other wanted proven horses, ready to work."[13] The population of horses on farms was made up largely of mares, unbroken horses and horses-in-training, while the urban horse workforce was made up largely of mature horses. As demand for horses to work in towns collapsed in the early decades of the twentieth century, large numbers of surplus town horses became available for use in agriculture and there was less incentive to breed replacement stock. As a result, the equine birth rate fell since the 1920s.

Felicity McWilliams argues that the long but dramatic decline in the population of agricultural horses distracts from, and even obscures, the considerable influence they continued to exert upon agricultural practices in Britain even after the introduction of the tractor.[14] Tractors were introduced into an agricultural context heavily shaped by horses and at first were used in conjunction with horses. Horses therefore continued to have some influence on the ways tractors were adopted and first used. Conventional accounts of the transition from horse to tractors have been simplistic and economically deterministic. McWilliams shows how for at least the first three decades of 'tractorisation', farmers defined good mechanisation not simply by the acquisition of the new technology but by how machines were used, and often by how they were used alongside horses. Farmers' continuing use of horses involved complex judgements about horses' individual characteristics, including breed, temperament, skill and work history, and their relationships with other farm workers, human and animal. Initially, the tractor was widely seen as complementary to the horse. Each technology had different strengths and weaknesses and by operating together each made the other more efficient.

The development of the tractor's pneumatic tyre gave the machine more flexibility and an ability to navigate a farm landscape that had been shaped by and for the horse. In the 1950s, tractors began to acquire new capabilities through power take-off and hydraulic implement mounts. McWilliams shows how it was only in the 1950s that the idea developed that tractors might eventually lead to a completely horseless future for farming. As late as 1958, the last year they were counted in the agricultural census, there were still 85,000 horses in use on British farms. The versatility of the horse had been seen as an advantage over the tractor, but the roles became reversed. Once this shift happened, and tractors became widely seen as the best way forward for draught power on farms, horses rapidly became cast as 'old-school' and traditional. Their use in rural land management withered to limited roles in forestry and woodland management, hauling cut trees over the more inaccessible terrain. Expert knowledge began to advocate not only the replacement of horses with tractors but the eradication of traces of horses from the agricultural landscape. Farm roads were repaved, modern new barns and sheds were constructed to house the new machinery, fields were enlarged and hedgerows were removed in order to better capitalise upon the efficiencies that mechanisation brought.

The dramatic transformation of the rural landscape from the 1950s to the 1970s prompted the growth of the British environmental movement and attracted considerable controversy during the 1980s as the scale of change began to be measured, understood and publicised. About 120,000 miles of hedgerows, around a quarter of the total, were removed in England and Wales, between 1946 and 1974, a radical change in the appearance of the rural landscape. However, this national aggregate figure, equivalent to some 4,500 miles of hedgerows removed each year, concealed local differences, and in some places the loss of hedgerows was even greater. In Norfolk, 45 per cent of hedgerows were removed in the 24-year period from 1946 to 1970.[15] This radical landscape change is usually put down to 'mechanisation' – the growth in the use of the tractor and combine harvester which worked more efficiently in larger fields. It can equally be thought of as the result of the end of the horse. As the horse disappeared from the agricultural landscape, so eventually did about half the hedgerows.[16]

Horses and Hunting

Hunting with horses and dogs predates the Norman Conquest, and English kings from the ninth to eleventh centuries were often revered for their hunting skills. King Harold is depicted in the Bayeaux tapestry riding to hounds, on one of 190 horses that feature in this embroidered record of the period. For most people in Anglo-Saxon times, hunting was carried out on foot with nets and using hunting dogs to chase and corral prey, and was principally for food rather than for pleasure. William the Conqueror was famed for his love of hunting deer on horseback, and he enjoyed regular hunting expeditions in the Forest of Dean. To protect his hunting, he developed a network of forests and a system of forest laws that have left "an indelible mark" on the ecology of the countryside.[17] Within tracts of land designated as royal forests, woodland was protected from tree-felling and from grazing animals that might destroy young trees. While previously people were free to hunt on their own land, under William the wild animals in the royal forests became the King's property. By the end of his rule, William had created 21 royal forests and his sons and successors consolidated and expanded the area of royal forests and firmly established hunting as a royal privilege. The New Forest was established, and the whole of Essex lay under forest law along with a large swathe of the Midlands from Lincolnshire to Oxfordshire. By the reign of Henry II (1154–89), the royal forest reached its greatest extent, covering almost a third of England's territory, all for the royal pleasure of hunting on horseback.[18]

Royal forests continued to be used through the twelfth to fifteenth centuries, and hunting on horseback served as a symbol of royal power and was esteemed as a preparation for warfare. Aristocratic hunting was accompanied by fine clothes, rituals, music and a special language, embellishments that helped enhance a sense of pageantry and status.[19] There were continual political struggles between the Crown and the aristocracy around forest land and hunting rights. By the fifteenth century, hunting a single male deer ('par force') using horses and hounds had become a popular form of the sport. This was a particularly thrilling type of hunting, involving unpredictability, chasing at speed, the risk of losing the trail and pitting the stamina and courage of the stag with the skills, guile and persistence of the hunt. It became the "premier pastime of all major landowners throughout the Middle Ages".[20]

After the Black Death wiped out between a third and a half of the English population in the late 1340s, pressure on resources eased and a golden age of hunting followed for a century until population growth began once more to increase pressures on the forests and woodlands established by the Normans. Deer parks were developed where tracts of land were fenced off to prevent deer from escaping. About 35 parks had existed at the time of the Domesday survey soon after the Norman Conquest, but many more were established and almost 2,000 have been identified during the Middle Ages. The largest medieval deer park, Clarendon Park, covered 4,500 acres (1,820 hectares) within which hunting seemed little different from that in open woodland.[21] This period of extensive rural landscape management for the purposes of managing deer as hunting quarry also saw the extinction of the wolf and boar from Britain. Tudor monarchs, especially Henry VIII and Elizabeth I, were enthusiastic hunters and took a keen interest in the management of forest lands and deer populations.

Over the 300 years from the 1350s to the 1650s, the English population trebled to around 6 million. Despite the civil wars and social unrest of the mid-seventeenth century, exclusive hunting rights of royalty and nobility remained intact. However, the deer population had been reduced to a fraction of its pre-Civil War levels. The Game Act of 1671 prohibited the killing of game except by qualified persons, who were restricted to the less than 1 per cent of England's population who were larger landowners, but effectively the gentry were given the right to hunt wild animals on anyone's land. Because foxes, martens, badgers, otters and wildcats lay outside the game laws, they could be hunted by smaller farmers and landowners, although they were not considered much fun to hunt at that time. Foxes were a farmyard pest and considered vermin, and, compared to deer, they ran relatively slowly and in straighter lines and so were not as exciting to hunt. They also tended to go to ground and had to be dug out. However, the history of hunting since Saxon times is one of adaptation to social and regulatory change as well as wildlife ecology, and in the late sixteenth century some hunters began to focus on how better sport could be derived from hunting the fox.

A set of developments centred in Leicestershire helped transform hunting with horses and hounds in the eighteenth century. Foxhounds were bred for stamina and speed, and the timing of hunting was shifted to mid-morning. Previously, foxes had been hunted in the early morning, but now foxes' earths were found and burrows blocked ahead of the hunt to prevent the quarry going to ground. Better quality and fast-running horses were bred, and the fine-tuning of the types and numbers of dogs and horses used, with the planning and orchestration of the hunt, ensured much better sport for hunters. A large part of the attraction of fox hunting was the speed and horsemanship required, the exhilaration and unpredictability of the fast ride and the difficulty in outwitting the fox. Foxhunting took off.

Until the early eighteenth century, hunting had largely been confined to the rural gentry. By the 1750s, the Quorn had become a fashionable hunt with its speedier hounds and by the 1780s, Quorn-style fashionable packs had been established across Leicestershire, Lincolnshire, Rutland and Northamptonshire, also the most prominent areas of horse breeding and trading.[22] By 1820, this form of hunting had been adopted all over the country. It became a key pastime for an element of fashionable society, but also began to involve a wider cross-section of farmers and other country dwellers who were able to follow the chase on horseback. As David Itzkowitz explains:

> By the beginning of the nineteenth century, foxhunting, which had in the past been a sport of merely local interest, had come to be seen as a national institution, as an integral part of country society, and as a bond among neighbours, irrespective of their social class. Hunting was, therefore, looked upon as a conservative force, which, by binding landlord, tenant and labourer together and giving them a sense of belonging to the same community, acted as a counterweight to the forces of radicalism, urbanisation, and industrialisation that threatened the stability of that community.[23]

By 1812 there were 69 packs of hunting hounds which increased to 91 by 1835 and 166 in 1866.[24] Hunting helped maintain a cohesive social fabric to rural life at a time when economic and technological changes were causing greater tensions between social classes. Members of all classes could mingle freely around the hunt and its associated social events and so landlord and tenant could feel they had a common interest. An etiquette developed and became institutionalised around hunters' access to land, and a system of hunt 'countries' developed with clear territorial boundaries between hunts. Riding to hounds could be expensive. Lord Althorp, who was master of the Pytchley from 1808, spent between £4,000 and £5,000 a year on the hunt. Early in the nineteenth century, Lord Fitzwilliam had spent £2,000–£2,500 a year on his ordinary riding horses and £500 more on his kennels. And when Sir Richard Burton retired as Master of the Burton Hunt in 1853, he kept about 100 horses and spent £10,000 a year.[25]

By the mid-nineteenth century, the development of the railways began to shape the geography of hunting and affect the pattern of participation. While railway lines cut across hunt countries and constrained where hunts could go, they opened the opportunity for hunting horses to be transported further afield and to join more distant hunting events. They also meant that participants could join hunts more readily from the towns and cities. However, opposition to hunting gathered pace in the second half of the nineteenth century as increasing numbers of people began to object to its unnecessary cruelty. As well as the cruelty to foxes, there was some concern about treatment of horses. An article in *The Times* on 12 March 1850 reported the death of a Mr George Darling in Northumberland who died after riding his horse to exhaustion.

> The horse which he was riding on was jaded to falling down, the run having been of the most terrible character ... The horse, being completely 'done', was standing still, in a deep furrow, with its head low, unable to move, when the rider struck the spurs into its sides, which caused it to spring and fall over in convulsive effort. The hind ridge of the saddle came with such weight and such a concussion upon the lower part of Mr. Darling's abdomen, as to cause the injuries which led to his death. ... We are compelled to speak in condemnatory terms of the furious way in which foxhunting has lately been pursued in Northumberland. It is assuming a shape of unmitigated cruelty; and if men set no value on their own lives, they ought to do so on those of their valuable animals. The hunt at which this melancholy accident occurred was absolute steeple chasing, and is denounced throughout the whole district. Some idea of its severity may be formed when we mention the fact that several horses died shortly after.[26]

Efforts to buy and sell foxes to address fox shortages in some areas gave lie to the claim that foxhunting was some essential form of pest-control. The League for the Prohibition of Cruel Sports was founded in 1924, and hunting with hounds became a totemic issue for animal rights campaigners for the rest of the twentieth century. Campaigning to ban hunting with hounds reached a peak in the 1980s and 1990s, and the election of the New Labour Government in 1997 on a manifesto offering a free vote on the question brought the issue to a head. Over 700 hours of parliamentary time was spent on the hunting issue.[27] A strategy of the hunting lobby was to seek to cast hunting with hounds as a significant activity in economic terms and claim a ban would threaten jobs and livelihoods in the rural economy.[28] At that time, claims abounded about levels of participation in, and expenditure on, hunting. For example, it was claimed that over 180,000 people participated in fox and deer hunting in Britain.[29] It was estimated that around 60,000 horses had been used for hunting at least once in a 1-year period (although a minority of these would have been kept solely for the purpose of hunting).

In 1999, amidst competing claims about the scale and significance of hunting wild mammals with hounds, the government established an independent Committee of Inquiry chaired by Lord Burns, former Permanent Secretary at the Treasury, to seek some authoritative clarity on the matter.[30] It found that there were 175 foxhunting packs registered in England and Wales, in addition to a small number of additional fell packs that did not use horses. There were around 15,000 hunt meets a year, with most hunts taking place twice a week. There were 67,000 hunt subscribers and supporters, the great majority of whom were assumed to be involved in foxhunting.[31] The Committee estimated that between 6,000 and 8,000 jobs depended on hunting with hounds. Of these, between 700 and 800 full-time equivalents were people directly employed by the hunts themselves, while a larger group, of 1,500 to 3,000 full-time equivalents, was employed by hunt participants, most commonly to care for hunting horses. The remaining employment was indirectly dependent on hunting in, for example, the manufacture of hunting clothing and equipment or in the hospitality sector. Following the Burns Committee's work, the government eventually did provide parliamentary time for a ban on hunting wild mammals with dogs, and hunts had to adapt their practices to hunt laid trails instead. The hunting ban did not bring an end to hunts, and it was not uncommon for hunts to report that after the ban, interest in participating and riding horses to hounds increased.[32]

Horses and the Rural Landscape

As we have seen, the human–horse relationship has shaped the rural landscape, not only through their work in agricultural labour, ploughing fields and hauling goods, and through the management of lands for hunting on horseback, but also through the extensive infrastructure required to feed and house them. Farm horses and horses for breeding required enclosed fields, usually bounded by fences, hedges or drystone walls. In addition to grazing, they would be fed hay and oats, so the upkeep of horses had implications for the kinds of crops that would be grown on farms. As towns and cities expanded, and the numbers of horses used in them increased, more land was required to grow the feed for working urban horses too. Annual consumption of oats and hay by town horses in Britain more than trebled between the 1830s and the end of the nineteenth century. Consumption of oats rose from 490,000 tons (440,000 tonnes) to 1,540,000 tons (1,400,000 tonnes), and consumption of hay rose from

840,000 tons (760,000 tonnes) to 2,640,000 (2,390,000 tonnes).[33] By the 1900s, fodder accounted for around 12 per cent of total British agricultural output. In the mid-nineteenth century, it was estimated that London required around 200,000 tons (181,000 tonnes) of hay a year to feed its horses, and specialised hay farms of the Home Counties devoted around 100,000 acres (40,000 hectares) of meadowland to helping service this need, although as much as a half of demand would have been met by hay production from further afield.[34]

Figure 5.1 shows the changing production of oats from the 1860s to the 1980s. The cultivated area under oats rose from 1.1 million hectares (2.7 m acres) in 1866 to over 1.3 million hectares (3.6 m acres) in 1895. The area under oats peaked in the First World War and again during the Second World War, but its rapid decline from the 1950s to the 1980s follows the decline in horse numbers over that period. As Figure 5.2 shows, for much of the late nineteenth century, oats were the most extensively grown cereal crop in Britain, but the proportion fell after the Second World War.

In addition to the impact of horse feed on agricultural cropping patterns, horses had to be housed and their feed had to be stored. Stables and barns were required as a standard part of farm infrastructure, and many remain as a prominent feature of the agricultural built environment. Until the eighteenth century, farming had largely been carried out using the open field system. About half of arable land had been enclosed by 1700, and subsequently, almost all agricultural land became enclosed.[35] The enclosures brought hedges, fences and drystone walling and a much more patterned and parcelled agricultural landscape. Enclosures also meant a loss of trees in thinly wooded countryside as vast quantities of timber were needed for fence posts and rails.[36] New by-roads

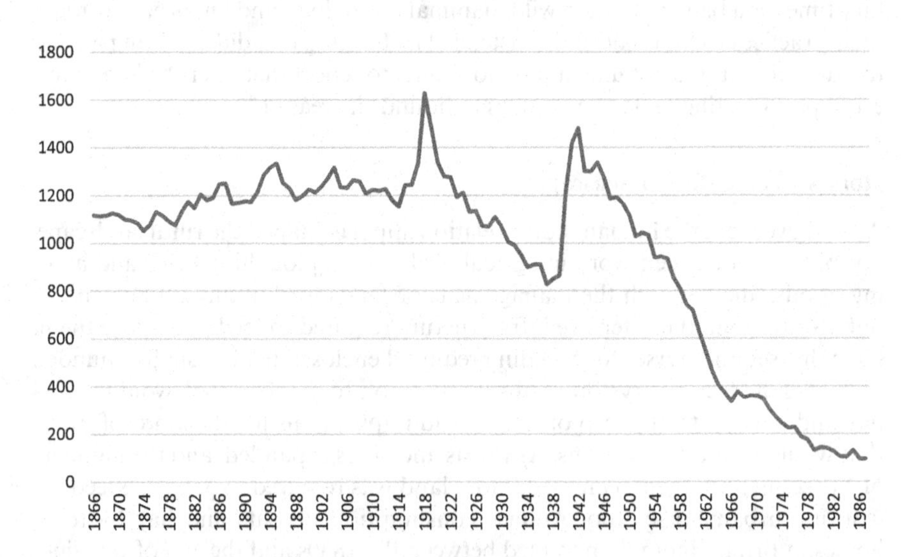

Figure 5.1 The Area of Oats Grown in Great Britain (000 ha), 1866–1987.

Source: Marks and Britton (1989), p.161. Data reproduced by permission of the Royal Agricultural Society of England.

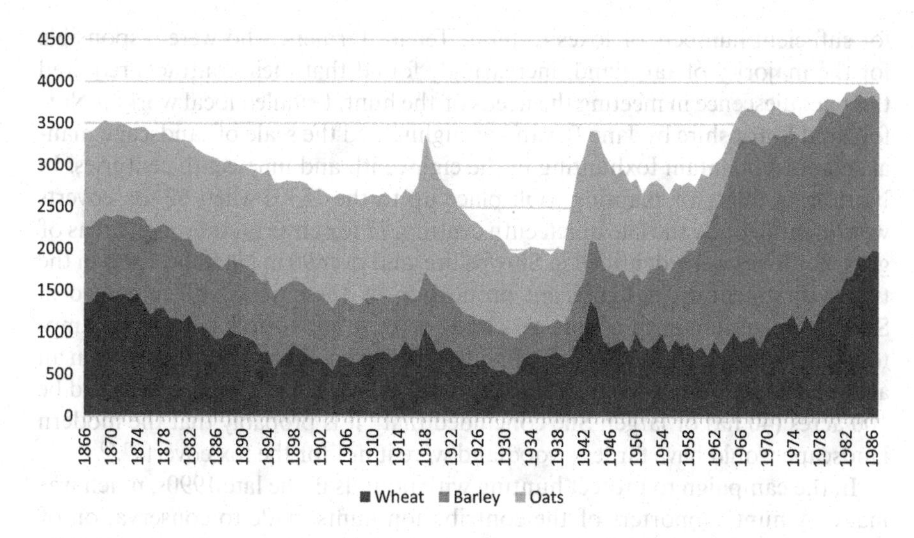

Figure 5.2 The Area of Cereal Crops Grown in Great Britain (000 ha), 1866–1987.

Source: Marks and Britton (1989), pp.158–61. Data reproduced by permission of the Royal Agricultural Society of England.

were established at the time of the enclosures to better link villages, wide enough to accommodate horse-drawn traffic. Forty-fee (12 metres) was widely regarded as the minimum width required for an unsurfaced road.[37] The enclosures also brought the need for farmsteads to be built beyond villages to enable the farming of enclosed land. Previously, under the open field system, farmworkers tended to live in nucleated villages, but the century from 1750 to 1850 saw isolated farmsteads built in the countryside beyond villages. Typically, these would accommodate the farm household, some farm workers, horses and other farm animals.

Between 1760 and 1825, the system of canals was developed, with a crucial role for horses in pulling boats and barges along the towpaths. By the end of the 1820s, some 3,000 miles of canals had been constructed, connecting the industrial towns and cities of the north, but also over the pastoral Midlands and south of England. Canals brought their own landscape and infrastructure with towpaths, lock-keepers' cottages and stables for canal horses.[38] Canal horses had to be fed well and regularly with high-energy food. Horse feed was prepared and available at provender stores all along the canal system. All regular stopping places, such as wharfs, warehouses and canal-side inns, had stabling for canal horses. Because canal horses could wear out horseshoes in 4 to 6 weeks, an infrastructure of blacksmiths and farriers was also required to sustain the horsepower of the canal system.

As well as the impact of working horses on the farmed landscape, horse-based leisure pursuits such as hunting also helped shape landscapes and underpinned landscape management. The enclosures initially made foxhunting more difficult, but gradually jumping hedges became incorporated into the hunt. Coverts – small blocks of woodland – were planted in order to provide habitat for sufficient numbers of foxes to hunt. Tenant farmers, who were responsible for the majority of farmland, increasingly found that their contracts required their acquiescence in meeting the needs of the hunt. Detailed local work in Norfolk and Shropshire by Jane Bevan has highlighted the scale of landscape management to maintain foxhunting in the eighteenth and nineteenth centuries. In Norfolk, a flurry of planting took place up to the 1830s when 62 fox coverts were identified. By the late nineteenth century, 77 fox coverts and rough areas of gorse for foxes were identified in Shropshire, and over 90 in Norfolk. Even in the twenty-first century, a significant proportion of these areas still remained in Shropshire, although three quarters had been lost in Norfolk.[39] In Northamptonshire, research by Mandy de Belin identified over 70 fox coverts planted in an area of 120 km^2. Many were around 20 acres (8.3 ha) in size, but some could be 100 acres (40 ha) or larger. She concluded that "it is probably that the modern landscape would have far less woodland were it not for the fox coverts".[40]

In the campaign to protect hunting with hounds in the late 1990s, much was made by hunt supporters of the contribution hunts made to conservation of valued landscape features. A survey by the British Field Sports Society in 1995 of over 150 lowland hunts, for example, found that 60 per cent of hunts carried out active covert management, laying shrubs and managing small woodlands to improve habitat for foxes and which also bring benefits for other forms of wildlife. This active landscape management can be directly attributed to the pleasures of horse-riding and hunting. Almost all hunts did work to repair and rehang gates, and over four-fifths did maintenance work on bridle paths. Over a fifth of hunts laid hedges.[41] The question of the positive contribution of hunting to the management and conservation of habitat was considered by the Burns Committee. It reported evidence that farmers who hunted were less likely to remove hedgerows and other non-productive habitat during the decades of agricultural intensification. It was, however, relatively unimpressed with the claims by hunting groups about the impact on woodland management. Some 5,700 hectares of woodland were initially reported to be managed by hunts, but the Committee calculated that this represented just 2.2 per cent of all woodland in England. Later work suggested that hunts could be managing around 10 per cent of the woodland available to them in their hunt countries.[42] The Committee found that "foxhunting has undoubtedly had a beneficial influence in lowland parts of England in conserving and promoting habitat which has helped biodiversity, although any effect has been in specific localities".[43] It also concluded that a ban on hunting with hounds would be unlikely to have a noticeable detrimental effect on landscape and conservation efforts, given the general direction of agricultural policy reform to increasingly support habitat management.

A key feature of the rural landscape associated with horses is the bridleway. These are established public rights of way, not only on foot but also on horseback or leading a horse. In 2021 it was calculated that horse riders in England had access to 22 per cent of all rights of way (some 42,000 km) comprising byways, restricted byways and bridleways. If unclassified roads were also included, the figure increased to over 57,000 km. Equestrian routes rely heavily on the historic network of highways and byways that are no longer part of the national network of public tarmacked roads. These tend to be old through-routes that run from parish to parish, linking one market town with another.[44] In 2002, the Countryside Agency, the then statutory agency responsible for public access to the countryside, estimated that there were some 16,000 km of unrecorded rights of way, a quarter of which (4,000 km) were bridleways and 2,700 km of which were byways open to all traffic. Maintenance of public rights of way is the responsibility of local authorities, but footpaths and bridleways have never been a significant and prestigious part of local government's work and can be a low priority in times of resource pressures. After the Second World War, many rights of way that had been blocked during the war were not reopened. Horse-riders were unaware of the need to record public rights of way, and many old bridleways and minor roads were not recorded or recorded only as footpaths. In the 30 years after the war, the network of recorded statutory bridleways started to shrink. From the 1980s, increasing interest in countryside recreation and in opening longer-distance bridleway trails raised hopes that the existing network could be expanded, and some local authorities established bridleway improvement schemes. The 1990s saw much greater use of the countryside for cycling, and sales of mountain-bikes boomed. The Labour Government legislated for improved public access to the countryside through its Countryside and Rights of Way Act 2000 which extended a so-called 'right to roam', but this greater access was only for those on foot. A cut off point of 1 January 2026 was established, beyond which rights of access would be lost on all those routes not officially registered and included on the 'definitive map' of rights of way.[45] Equestrian bodies have therefore been struggling to protect rights to ride horses at different sites and along particular paths.

Horses themselves can be thought of as a feature of the rural landscape and are prominent in paintings of countryside landscapes. Constable's *The Hay Wain*, painted in 1821, is one of the most popular examples of rural landscape painting and depicts three horses pulling a wagon across a river in East Anglia. Paintings and drawings of hunting scenes featuring horses galloping across the countryside are a stock feature of country houses and country pubs across the land. Richard Yarwood and Nick Evans have traced how rare breeds of livestock animals have become important signifiers of local and regional distinctiveness. Just as architectural features of the industrial past have become commodified as part of the gentrification and re-imagination of docksides, waterfronts and the cultural quarters of post-industrial cities, so there has been a similar process with respect to distinctive farm livestock in the countryside.[46] Farm horses and rare breeds commonly feature in farm tourism and in rural heritage museums. The survival of rare and distinctive breeds of livestock depends on the attitudes of people towards them. As parts of British agriculture become oriented to farm tourism, rural leisure and wildlife conservation, farmers and land managers become landscape and wildlife curators and rare breeds, including breeds of farm horses, become part of the landscape they are curating.

Horses and the Twenty-First-Century Consumption Countryside

Over the last 40 years, there has been a marked shift in the social and economic complexion of the countryside and in the prevailing concerns about how rural land and space is managed. This shift is typically understood as one where concerns about the production of food and fibre have been replaced by a growing focus on the quality of the rural environment as a site for nature conservation, leisure and recreation, a shift in emphasis from production to consumption.[47] During the second half of the twentieth century, countryside policy was dominated initially by a national imperative to expand agricultural production after the Second World War. The wartime experience of vulnerability of food supplies and the 'dig for victory' campaign inspired a very different approach to rural land than that which had prevailed over the long period from the repeal of the Corn Laws in 1846 to the outbreak of the Second World War. During that period, a laissez-faire cheap food policy was pursued, and food was largely imported from overseas. British agriculture was not financially supported, and farmers had to fend for themselves competing in world markets. There was little investment in agriculture, and a long agricultural depression meant declining farm incomes and a general exodus from the land. During the Second World War, a radically new approach was devised which sought to support and protect British farmers and, through stimulating a technological revolution in agriculture, improve agricultural productivity and make best and most productive use of the British land resource.

Food production enjoyed protected and privileged status over the 'productivist' period from the 1940s to the 1980s, but from the 1980s onwards, the policy model came under sharp criticism and began to be reformed. Surplus production built up in the 1980s and there was increasing evidence of environmental problems associated with intensive farm production practices. Various measures were introduced to curb over-production, better control the cost of agricultural support and grapple with modern farming's growing range of environmental problems such as damage to valued wildlife habitats and pollution of water. Alongside food production, rural land became increasingly valued for its biodiversity and environmental services, for its aesthetic and amenity value and as a site of recreation. Agricultural policy reforms sought to help farmers to adjust and find new sources of income. Social and economic changes also altered the use of rural spaces. More middle-class people moved out of larger cities into smaller towns and villages in more rural areas.[48] Where through the first half of the twentieth century the dominant issue in rural geography was one of population loss and rural decline, from the 1970s onwards waves of affluent 'counterurbanisers' were transforming the social complexion of the countryside.[49] More affluent in-migrants moved not only into the attractive larger villages and small towns in accessible rural areas, but were increasingly buying homes, often converted from former farm buildings, in the deep countryside. New residential properties were developed from redundant farm buildings, which included stables and barns previously used to house and service agricultural horses.

The late twentieth century also saw increasing participation in countryside recreation. By the 1990s, around 25 per cent of the population reported to be frequent visitors to the countryside with a further 50 per cent occasional visitors. Those participating in countryside recreation were more likely to be families from more affluent social classes.[50] Leisure and consumption practices generally became more important influences upon the social and cultural identities of individuals and groups. This was part of a wider social trend in which national society evolved from being organised around industrial production to a post-industrial society. Social identity, rather than being something ascribed and fixed, became more flexible and chosen.[51] As self-identity shifted to spheres where individuals had more control, a growing focus on consumption and leisure activities developed.[52] The 1980s and 1990s saw a growth in a range of outdoor leisure pursuits such as mountain-biking, paintballing and birdwatching. The visitor economy became an increasingly important part of local rural economic development, and a countryside leisure economy developed alongside the traditional agricultural economy. The development of the horse-based economy worked very much with the grain of prevailing social and market trends in the late twentieth century. Squeezed by subsidy reforms, farmers looked for new sources of income. Developing liveries and horse-based leisure and tourism was one popular farm diversification strategy. In a study of rural change in Buckinghamshire in the early 1990s, Jonathan Murdoch and Terry Marsden found that a quarter of farmers in their survey had either taken land out of agricultural production or were planning to do so. Farmers were establishing liveries and providing land and buildings to graze and stable horses. They were establishing farm shops and tourist attractions, stocking ponds with fish to let for angling, converting farm buildings into offices and light industrial units. Farmers were moving from being conventional food producers to being highly diversified rural entrepreneurs in the vanguard of developing the countryside as a place of leisure and consumption.[53]

The diversification of farming enterprises was actively supported by governments across the European Union and in the UK. Farming organisations were initially hesitant about adopting farm diversification, which had been seen as something that 'failing farmers' had to resort to, but social and economic trends and the direction of policy support meant diversification became more and more widespread. In a report on the farming economy in 1999, the National Farmers' Union (NFU) reported that 62 per cent of farms received some income from sources outside farming, and over the previous decade, over a fifth of total farm incomes had come from sources other than agriculture.[54] Even among full-time farms, those traditionally most strongly focussed on food production, significant numbers were diversifying. The Ministry of Agriculture, Fisheries and Food (MAFF) reported that in 1998/99 the proportion of full-time farms in England with an on-farm diversified enterprise was 25 per cent, with the proportion being highest among smaller farms (29 per cent) and among farms in the west (32 per cent).[55] An NFU survey found that letting property was the most common form of farm diversification, followed by haulage and construction enterprises, with keeping and breeding horses the third most common. Some 12 per cent of farmers had a horse enterprise.[56]

Over a period of a century or so, this shift from production to consumption can be mapped through the changing uses and roles of the horse in the countryside. In 1911, there were over 1.25 million horses used on English farms, plus a further 163,000 in Wales, the vast majority of which were used for solely agricultural purposes.[57] By 1940, numbers had almost halved and by the 1960s, numbers were so small that they were no longer included in agricultural statistics. However, the numbers of horses used for leisure steadily rose after the 1980s, and by 2010 there were around a million horses in the UK, the vast majority of which were used in rural areas and for leisure purposes. The horse economy is one of the most extensive land uses in England after agriculture, forestry and shooting. This includes the grazing paddocks for horses attached to houses and farms in the countryside, stables and stud farms and horse-racing and related land uses.

Research by Lee-Ann Sutherland in rural Scotland has considered the role that a commitment to equestrianism can play in residential location decisions and in the gentrification of rural landscapes.[58] She presents a pattern of small-scale horse-keeping as a key trend in rural land use change across northern Europe, especially in peri-urban areas. Research in Belgium found that one third of agricultural land in a study area in Flanders was used for recreational horse-keeping.[59] Similar rapid growth in horse-keeping was identified in the accessible rural land around Berlin.[60] Sutherland uses the term 'horsification' to describe a form of rural gentrification specifically to enable horse-based recreation. Horsification is often heavily gendered with the establishment of a new household of humans and horses, usually because of women's commitment to horse-keeping. She studied 16 horse-keepers and found that keeping horses had been the primary reason for relocating to, or within, the countryside in all cases. Capital investment took place in smallholdings to ensure they were suitable for keeping horses, and this increased the value of the properties but also reshaped the local rural landscape and altered the local ecology (and not always for the better).

> In collaboration with their humans, horses territorialise a range of public and private spaces, not limited to the installation of new fencing, stabling and other facilities on their residential properties. Horse bodies "claim" land through grazing it – pasture fields become paddocks. The land itself is enrolled into the gentrification processes through the agencies of flora and fauna, and the identities which are performed in relation to it.[61]

The rise of recreational horse-keeping helps create an equestrianised rural landscape which, over the past 40 years, has become an increasingly prominent feature of the British countryside and even a dominant one in some places closer to urban settlements. In addition to the landscape developed by individual recreational horse-keepers is that associated with riding schools, liveries and farm-based equestrian centres, which have all contributed to the shift from a productivist to a post-productivist rural leisure landscape.

Conclusions

This chapter has traced the ways that the uses of horses have shaped Britain's rural geography. Prominent among these is the use of the horse as an agricultural draught animal which came to dominance over the fifteenth and sixteenth centuries. The gradual replacement of the oxen with the horse was a technological change with a geography of its own, influenced by rainfall, soil conditions and cropping patterns. By the seventeenth century, horses made up over two-thirds of agricultural draught animals and dominated agricultural 'horse-power', to which they gave their name, for three centuries until the mid-twentieth century. Their use helped speed up ploughing and harrowing, which may have helped improve the productivity of the land but certainly saved farm labour for other purposes. They brought significant benefits in haulage and helped geographically extend the operation of local rural markets for goods. They played their part in developing the preconditions for the agricultural revolution which was to transform British agriculture, then industry, settlement structure and the nation's economic and political power and its place in the world.

Alongside this pivotal role in agricultural work, horses continued to be at the centre of the leisure pursuits of the aristocracy and landed classes. Royal passion for hunting on horseback had ultimately resulted in huge swathes of England, peaking at a third of all land, coming under forest law to ensure quarry and forest resources were managed in ways that protected the sovereign's sport. As par force hunting on horseback evolved through the centuries in response to social, ecological and regulatory change, so the practice became part of an intricate web of social relations through which the notion of a cohesive rural community was performed. Hunting on horseback came to be seen, and defended, by its protagonists as the very epitome of British rural tradition. The rural landscape was increasingly arranged and managed to provide good sport alongside its food production functions. The countryside became configured around the pleasures to be derived from horse-backed sport and recreation.

At the turn of the twentieth century, a third of Britain's 3.3 million horses were used in agriculture. Most horses were powering an urbanising and industrialising economy, hauling people and things from place to place. As the century progressed, the uses of horses in towns and cities fell away quickly. From the 1920s onwards, most horses were to be found in the countryside, although the numbers in the countryside dropped too as the century progressed. By the second half of the century, total horse numbers had fallen by four-fifths. With the exception of police horses, the royal household cavalry and the odd draught horse pulling wagons for brewers, coal merchants and rag-and-bone men, horses were essentially now creatures of the country. However, these were no longer animals on the decline, fading relics of a past equestrianised agricultural age. They were becoming an important and increasingly popular force in the transition to the post-productivist consumption countryside. As agricultural change brought a new age of the diversified farm enterprise, farmers moved from being simply food producers to be multifunctional rural entrepreneurs. They developed farm buildings as attractive new homes for affluent incomers. They became landlords of rural industrial units. They stocked ponds to let for fishing, and they played their role in developing the new horse economy as providers of livery yards and grazing land for horses. Land and accommodation for horses became a major rural land use over the last two decades of the twentieth century.

Equestrian businesses are a quintessential part of the rural economy, and proudly proclaim this role for themselves. Owning and looking after horses has become common and widespread in the consumption countryside, and equines continued to shape the rural landscape. The manicured, fenced, wooded and secluded settings of stud farms might represent the physical expression of the elite end of the horse economy in the rural landscape. Much more common, however, are the fenced paddocks, with their little wooden shelters and crossed poles for horse-jumps, scattered extensively across rural areas, and managed by the 330,000 households who own horses.

Notes

1 McWilliams, F. (2019) *Equine Machines: Horses and Tractors on British Farms c.1920–1970*. King's College London PhD Thesis, Department of History.
2 Piggott, S. (1981) Early prehistory, pp.3–62 in S, Piggott, S. (ed.) *The Agrarian History of England and Wales Volume I-I Prehistory*. Cambridge: Cambridge University Press, p.58.
3 Langdon, J. (1986) *Horses, Oxen and Technological Innovation: The Use of Draught Animals in English Farming 1066–1500*. Cambridge: Cambridge University Press.
4 Langdon (1986), p.29.
5 Fraser, T. (2019) *Livestock and Landscape: Livestock Improvement and Landscape Enclosure in Late and Post-Medieval England*, PhD Thesis, University of Sheffield, Department of Archaeology, pp.315–16.
6 Kent, N. (1794) *Agricultural Survey of Norfolk*, p.141, quoted in Riches, N. (1967) *The Agricultural Revolution in Norfolk*. (Second Edition). London: Frank Cass, pp.106–7.
7 Holderness, B. (1989) Prices, productivity and output, pp.84–274 in G. Mingay (ed.) *The Agrarian History of England and Wales, Volume VI, 1750–1850*. Cambridge: Cambridge University Press, p.131.
8 Thompson, F.M.L. (1976) Nineteenth century horse sense, *Economic History Review* 29, p.80.
9 Barker, R. (1983) The delayed decline of the horse in the twentieth century, p.101–112 in F.M.L. Thompson (ed.) *Horses in European Economic History: A Preliminary Canter*. Reading: British Agricultural History Society.
10 Collins, E. (1983) The farm horse economy of England and Wales in the early tractor age 1900–40, pp.73–100 in F.M.L. Thompson (ed.) *Horses in European Economic History: A Preliminary Canter*. Reading: British Agricultural History Society, p.73.

11 Collins (1983).
12 Collins (1983), p.85.
13 Collins (1983), p.87.
14 McWilliams (2019).
15 Baird, W. and Tarrant, J. (1973) *Hedgerow Destruction in Norfolk, 1946–1970*. Centre of East Anglian Studies. Norwich: University of East Anglia.
16 Shoard, M. (1980) *The Theft of the Countryside*. London: Temple Smith, p.34.
17 Griffin, E. (2007) *Blood Sport: Hunting in Britain Since 1066*. Yale University Press, p.15.
18 Hoskins, W. (1955) *The Making of the English Landscape*. Harmondsworth: Penguin, pp.90–91.
19 Griffin (2007), p.31.
20 Griffin (2007), p.56.
21 Griffin (2007), pp.63–5.
22 Porter, J. (1989) The development of rural society, pp.836–937 in G. Mingay (ed.) *The Agrarian History of England and Wales Volume VI 1750–1850 Part II*. Cambridge: Cambridge University Press, p.919.
23 Itzkowitz, D. (2016) *Peculiar Privilege: A Social History of English Foxhunting, 1753–1885*. Brighton: Edward Everard Root, p.x.
24 Porter (1989), p.920.
25 Porter (1989), pp.920–21.
26 Quoted in May, A. (2013) *The Fox-Hunting Controversy, 1781–2004: Class and Cruelty*. Farnham: Ashgate, p.65.
27 May (2013), p.1.
28 Ward, N. (1999) Foxing the nation: the economic (in)significance of hunting with hounds in Britain, *Journal of Rural Studies* 15, 389–403.
29 Cobham Resource Consultants (1997) *Countryside Sports - Their Economic, Social and Conservation Significance*, Reading: Standing Conference on Countryside Sports.
30 Committee of Inquiry into Hunting with Dogs in England and Wales (2000) *Report of the Committee of Inquiry into Hunting with Dogs in England and Wales* Cm.4763. London: Home Office.
31 Committee of Inquiry into Hunting with Dogs in England and Wales (2000), p.34.
32 Harvey, F. (2013) Huge rise in first time hunters for new foxhunting season, *The Guardian* 2nd November, p.18.
33 Thomson, F.M.L. (1983) Horses and hay in Britain 1830 to 1900, pp.50–72 in F.M.L. Thompson (ed.) *Horses in European Economic History: A Preliminary Canter*. Reading: British Agricultural History Society, p.60.
34 Thompson (1983), p.65.
35 Hoskins (1955), p.178.
36 Hoskins (1955), p.196.
37 Hoskins (1955), p.202.
38 Hoskins (1955), p.253.
39 Bevan, J. (2011) *Foxhunting and the Landscape between 1700 and 1900 With Particular Reference to Norfolk and Shropshire*. PhD Thesis, University of East Anglia, School of History.
40 de Belin, M. (2013) *From the Deer to the Fox: The Hunting Tradition and the Landscape 1600–1850*. Hatfield: University of Hertfordshire Press, p.92.
41 Cobham Resource Consultants (1997), pp.99–100.
42 Committee of Inquiry into Hunting with Dogs in England and Wales (2000), p.126.
43 Committee of Inquiry into Hunting with Dogs in England and Wales (2000), p.131.
44 Francis-Baker, T. (2023) *The Bridleway: How Horses Shaped the British Landscape*. London: Bloomsbury.

45 Bucks, S. and Wadey, P. (2017) *Rights of Way: Restoring the Record* (Second edition) Ilminster: Bucks & Wadey Publishing.
46 Yarwood, R. and Evans, N. (1998) New places of 'old spots': The changing geographies of domestic livestock animals, *Society and Animals* 6, 137–65.
47 Marsden, T. *et al.* (1993) *Constructing the Countryside*, London: University College London Press.
48 Marsden, T. *et al.* (1993).
49 Lowe, P. *et al.* (1995) *Countryside Prospects, 1995–2010: Some Future Trends*, CRE, University of Newcastle, p.3.
50 Lowe *et al.* (1995), pp.22–4.
51 Lash, S. and Urry, J. (1994) *Economies of Signs and Space*. London: Sage.
52 Urry, J. (1995) *Consuming Places*. London: Routledge.
53 Murdoch, J. and Marsden, T. (1994) *Reconstituting Rurality: Class, Community and Power in the Development Process*. London: UCL Press.
54 National Farmers' Union (1999) *Routes to Prosperity for UK Agriculture*. London: National Farmers' Union, p.60.
55 Ministry of Agriculture, Fisheries and Food (2000) *Current and Prospective Economic Situation in Agriculture*. Working Paper. London: MAFF.
56 NFU (1999), p.68.
57 Collins, E. (2000) *The Agrarian History of England and Wales Volume VII 1850–1914*, Parts I & II. Cambridge University Press, pp.1179–84.
58 Sutherland, L.A. (2021) Horsification: Embodied gentrification in rural landscapes, *Geoforum* 126, 37–47.
59 Bomans, K. *et al.* (2011) Pasture for horses: An underestimated land use class in an urbanised and multifunctional area, *International Journal of Sustainable Development and Planning* 6, 195–211.
60 Zasada, I. *et al.* (2013) Horsekeeping and the periurban development in the Berlin Metropolitan Region. *Journal of Land Use Science* 8, 199–214.
61 Sutherland (2021), p.45.

6 The Horse Economy

Introduction

Globally, the equestrian industry is estimated to have an economic footprint of some £250 billion. It is described by Christina Jones, founder of the Equine Business Association, as "one of the biggest industries that is hidden in plain sight".[1] Around a third of the global industry is based in the United States, while over two-fifths are in Europe. Globally, the horse industry supports an estimated 1.6 million jobs, of which 460,000 are in North America and 400,000 are in Europe.[2] The European Horse Network estimates that there are around 7 million horses in Europe and that around 2.6 million hectares of permanent pasture are given over to horse grazing.[3]

Taking a global perspective, most of the world's 60 million horses are working horses. In the UK, the horse economy is now dominated by horses used for sport or recreation. Recreational horses, which are the large majority, may be owned by either private individuals or horse-riding businesses. These are ridden principally just for leisure or may participate in unaccredited equestrian events. In addition to these are sporting horses such as racehorses and show-jumping horses which are ridden principally in competitive sport. A final category of working horses includes police horses and the military's ceremonial horses along with small numbers that work in agriculture and forestry, pull carriages for weddings and funerals or work at visitor attractions. With no authoritative census of horse numbers, the size of the horse economy is dependent upon surveys conducted every few years for the industry to provide estimates of horse numbers and the scale of economic activity. It is difficult to be precise, but we can say that the current population of horses is likely to be under a quarter of the peak level in the early twentieth century. There is some evidence to suggest that horse numbers grew through the 1990s to a more recent peak of just over 1 million in 2005, but numbers have since steadily fallen back to around 750,000 in 2023. However, while horse numbers may have declined over the last decade or so, the number of people participating in horse-riding and related activities had increased up to the Covid pandemic and the economic crisis that followed.

DOI: 10.4324/9781003454359-6

The total size of the horse economy depends upon the definition that is used. The surveys commissioned for the British Equestrian Trade Association (BETA) provide a standardised means of estimating the economic value of the equestrian sector every few years in terms of private and professional spending on equestrian goods and services. This is estimated to be £5 billion in 2023 and has grown from £3.4 billion in 2006, £4.3 billion in 2015 and £4.7 billion in 2019. However, it does not include spending by horse-racing spectators and by betting on horseracing.

BETA's 2023 £5 billion estimate is based on the economic activity associated with 630,000 privately owned horses and 95,000 owned by professional equine establishments associated with the horse-racing industry.[4] The British Horse Industry Confederation estimated an annual turnover of over £7 billion for the horse industry in Great Britain in 2010.[5] More recent estimates in 2019 suggested that the horse industry had an annual turnover of approximately £8 billion, which included gambling on horseracing.[6]

This chapter examines the recent evolution of the horse economy. It traces the changing economic context for the horse industry during the late twentieth century, focusing on the growth in uses of horses for sport and leisure. It examines the moves to define and develop a unified 'horse sector' over the past two decades, and reflects on the contemporary geography of the sector before considering those places where the horse industry is particularly prominent, such as Newmarket, the largest racehorse training centre in the UK and seen by many as a global centre for horse breeding and horse health.

The Horseracing Economy

As we saw in Chapter 4, horseracing has been practiced as a sport for many centuries. The Romans raced chariots, and racing horses became an elite pastime of the aristocracy.[7] Horseraces took place in medieval times, often just between two horses, and began to be more organised and formalised as a spectator sport from the sixteenth century onwards. Horseracing became the first and longest-lasting British national sport, and the eighteenth century was a key period in its development. Race meetings developed from marginal and informal events among the elite to become among the most important leisure events of the summer season and an important part of the general commercialisation of leisure.[8]

Horse racing is now a globalised industry with strong representation in the UK, the United States and Australia but with participation across a range of other countries. Around 90,000 thoroughbred foals are born around the world each year, and thoroughbred horses are bred and traded within a globalised network of relationships. The two main drivers of the industry's development are the consumption of racing as a betting medium and the production of thoroughbred racehorses. The sport is extensively televised and often commands relatively high viewing numbers. In the UK, horseracing is the second largest spectator sport after football.[9] British horseracing is largely on grass and can either be flat racing or over jumps. (In the United States, most racing is on dirt rather than grass, and without jumps.) National hunt racing in Britain evolved in association with hunting, with horses racing over a course with jumps and ditches.

Thoroughbred racehorse breeding is carefully managed through a system of studbooks that trace the lineages of all registered thoroughbred horses. There are 70 recognised national stud books on the international stud book list, including Argentina, China, Denmark, India, Morocco and South Africa.[10] Thoroughbred breeding originated in Britain in the seventeenth century, and a small number of stallions brought to Britain helped establish the thoroughbred breed. Three imported stallions, the Byerley Turk (1680s), the Darley Arabian (1704) and the Goldolphin Arabian (1729), are said to have been particularly influential in establishing the bloodlines that were codified in the General Studbook. Genetic research has found that the Y-chromosome of 95 per cent of modern racehorses can be traced back to a single horse, the Darley Arabian.[11] Thoroughbred breeding is continually being refined to breed racehorses for speed, and the work of the International Stud Book Committee since 1976 has ensured increasing international standardisation of horse pedigree management and governance. Only physical breeding by mounting is permitted and artificial insemination is prohibited, so international breeding requires the international movement of thoroughbred horses.[12] This has resulted in the development of a global network of thoroughbred breeding clusters including Newmarket in England and Lexington in Kentucky.[13]

Britain's thoroughbred industry is recognised globally for its world-class breeding, training and racing. A top breeding stallion can attract a fee of several hundred thousand pounds per mare and can earn tens of millions of pounds in a year. The industry accounts for over £375 million of economic activity in the UK and supports over 19,000 jobs, with most of the economic impacts focussed in the rural economy.[14] Breeding underpins a supporting equine service sector which includes auction houses, racecourses, racing media and gambling. The UK produces a higher proportion of the world's top ranked thoroughbred horses of any other country and hosts 24 of the top 100 rated horseraces.[15] According to an analysis of the industry in 2018,

> Great Britain's favourable breeding environment, high standard of equine welfare laws, excellent infrastructure, and highly skilled staff make it one of the best countries in the world in which to establish a breeding operation. As a result many global breeders locate in Great Britain, injecting substantial foreign investment into the economy and producing well-paid jobs both in breeding and supporting industries.[16]

The size of the British horseracing industry has not grown significantly over the last decade.[17] The racing economy is expensive to sustain and depends upon wealthy racehorse owners being prepared to pay for the upkeep and training of their horses. The average price of a yearling racehorse sold at Tattersalls in Newmarket in 2018 was 150,000 guineas.[18] (The most expensive racehorse ever,

Fusaichi Pegasus, was sold in the United States for almost £54 million in 2000.) It costs around £45,000 per year to keep a racehorse in training. Typically, owners can expect on average to get back only £21 of every £100 spent. There are currently around 14,000 racehorses in training in the UK at a net cost to their 8,000 owners of £500 million to £600 million a year.[19] (Rebecca Cassidy in 2020 reported a Racehorse Owners' Association estimate that it cost £20,000 a year to keep a racehorse in training, excluding entry fees, vet bills and insurance[20] but by 2023 the cost of keeping a top racehorse in training was estimated at more than twice that).

Attendance at horseracing events has been relatively stable over recent decades, despite the disruptions of Foot and Mouth Disease (FMD) in 2001 and Covid in 2020/21. A new terrestrial TV deal came into effect from 2013 which provided some financial stability. In 2019, 5.62 million people attended over 1,500 individual race meetings in England, Scotland and Wales, with the number of fixtures remaining fairly steady over the previous decade. Key horseracing events can attract up to 1.5 million live TV viewers. In 2020, the sector was estimated to generate some £4.1 billion of economic activity, with over 20,000 people employed on 59 licensed racecourses around Britain. Around 45 per cent of racecourse revenue is generated by attendees at racing events.[21]

The sector suffered an economic squeeze in the years before the Covid crisis. Breeders increasingly struggled to break even, and the proportion of breeders operating at a loss increased from 45 per cent in 2013 to 66 per cent in 2017. The squeeze led to a polarisation of the economic prospects, with smaller breeders being the worst affected. Between 2013 and 2017, 350 breeders exited the industry.[22] There has also been a growing influence of Irish national hunt horses in British racing, with over 50 per cent of national hunt horses in training Irish bred and only 33 per cent British bred. Thoroughbred imports have grown faster than exports, such that the trade gap has doubled since 2012.[23] The sector has suffered from staffing shortages and skills gaps, and the new frictions and costs in the international movement of horses since Brexit have eroded margins for breeders. Prior to Brexit, there were around 27,000 horse movements between the UK, Ireland and France every year, making up 87 per cent of total horse movements across the whole of the EU.[24]

The horseracing industry is dependent upon the gambling economy. As Richard Blackmore put it in 1891, "Betting is the manure to which the enormous crop of horseraces and racehorse breeding is to a large extent due".[25] Betting on racing grew significantly in the nineteenth century but became a concern to governments in the twentieth century and so was regulated and restricted. Off-course betting on horseracing was legalised in 1961 which permitted the development of betting shops, and a Horseracing Betting Levy helped fund the horseracing sector.[26] Today, there are over 8,400 betting shops in Britain. A proportion of the profits generated by betting companies are taxed and distributed back to the horseracing industry in the form of prize money and grants. The introduction of the National Lottery in 1994 and the 2005 Gambling Act opened up the betting world to new types of betting products and there has also been marked growth in fixed odds betting terminals (including fruit machines and other electronic betting games).

Gambling remains big business in the UK and around the world. In 2015 the global gambling industry was estimated by Morgan Stanley to be worth $423 billion (£344 billion) and was predicted to grow to $635 billion (£516 billion) by 2022.[27] The annual market value of gambling in the UK was estimated to be around £15 billion a year between 2017/18 and 2019/20. The Covid crisis meant gambling turnover fell to £12 billion in 2020/21 but then recovered to £14.7 billion in 2021/22. Mintel forecast that the UK market is likely to continue to grow, with its best estimate forecast of £18.8 billion by 2026/27. Only a quarter of the total gambling market is made up of conventional betting. Some £4.3 billion (or 30 per cent) comes from gaming, £4.0 billion (28 per cent) from lotteries and £3.8 billion (26 per cent) from betting. In the betting market, horseracing is losing market share as more betting is carried out online and focuses on football. Racing's position is stronger in betting shops, but even here there has been an increasing emphasis on gaming machines. Betting on horses has increasingly become seen as an 'old man's hobby', and young gamblers are more likely to focus on football.[28]

In 1998, horseracing accounted for 70 per cent of gambling in betting shops, but this fell to less than half by 2007 and only 30 per cent by 2014.[29] There has been much more active marketing by betting companies for betting on football, with more live betting, where sporting events like football matches are split into smaller events and gamblers are able to bet continually during the game on half-time scores, who scores and so on, rather than just the eventual winner. By 2015, 80 per cent of bet365's sports betting revenue came from betting in play.[30] Statutory Levy receipts for horseracing declined from a high point of £117 million in 2007/08 to £60 million in 2014/15. The arrangements were reformed in 2015, and the flow of money from the Horserace Betting Levy increased by £45 million in 2016/17.[31] However, the yield from the Levy has been volatile, falling from £94.7 million in 2017/18 to £83 million in 2018/19, before rising again to £97 million in 2019/20. The British Horseracing Authority complains that such volatility makes it difficult to plan investment.[32]

The Wider Horse Economy

Beyond racing are a range of equestrian sports, including the Olympic sporting disciplines of show-jumping, dressage and eventing, as well as non-Olympic sports such as horseball, reining, equestrian vaulting and endurance. Equestrian competition was first introduced to the Olympics in 1912 in the disciplines of dressage, showjumping and eventing. Competition was initially dominated by cavalry officers, and civilians did not win significant numbers of Olympic medals until the 1950s. The governing bodies for the various equestrian sports in Britain are members of British Equestrian (formerly the British Equestrian Federation), which is the UK's single member of the international governing body, the Fédération Équestre Internationale (FEI). British Equestrian has 18 different member associations. The largest is British Dressage (with around 17,500 members), followed by British Showjumping (15,000 members) and British Eventing (15,000 members).[33] Dressage is a competitive display of the disciplined control of horse movement and has its origins in elite training in military horsemanship in early modern times. Between 1750 and 1900, it became an advanced discipline. Showjumping has its origins in foxhunting and the competitive challenge of negotiating fences, gates, ditches and hedges. Eventing comprises the three disciplines of dressage, cross country and showjumping. Equestrian sports have become popular for television, and regular international competitions are held around the world.

Most horses are owned, usually in ones or twos, by private individual owners who look after them and ride them for pleasure. Most of these recreational horses will not be participating in sporting events. Regular national surveys for BETA provide estimates of the changing size of this large part of the horse economy. The survey is based on telephone and online surveys of a sample of the British population, supplemented by more detailed survey research among horse-related businesses. Survey results have been published in 1999, 2006, 2011, 2015, 2019 and 2023 (Table 6.1). The results are dependent upon the survey technique, are subject to sampling error and have moved around between surveys, so they have to be interpreted with some caution. The 2006 results seem to be high and outlying, but given a relatively standardised approach, the trends suggest that over the last decade more people ride horses, but the horse population is diminishing and fewer households are owning horses.

The results published in 2006 involved a representative survey of 5,078 British adults. It estimated that the population of horses owned by private individuals was 1.2 million, with a further 100,000 owned by professional establishments. This total estimated figure represented an unlikely 50 per cent growth in the number of privately owned horses between 1999 and 2005. Five years later, the results of a survey of a similar number of adults suggested that the total horse

Table 6.1 The Size of the British Horse Industry in the Twenty-First Century

	1999	*2006*	*2011*	*2015*	*2019*	*2023*
Estimated horse population (000)	900	1,300	988	944	847	726
Estimated number of regular* riders (million)	1.4	2.1	1.6	1.3	1.8	1.82
Estimated number of riders (in the last year) (million)	2.4	4.3	3.5	2.7	3.0	3.2
Estimated number of horse-owning households (000)	—	720	451	446	374	331

Source: National Equestrian Surveys for the British Equestrian Trade Association. Reproduced by permission of the British Equestrian Trade Association. * Regular riders are defined as those who ride at least once a month.

population was 988,000, a figure more in line with the general trends since 1999. Since 2011, three further surveys estimated a gradual decline in the number of horses to 944,000 in 2015, 847,000 in 2019 and 726,000 in 2023.[34] Similarly, there has been a decline in the number of households owning a horse from 451,000 in 2011 to 446,000 in 2015, 374,000 in 2019 and 331,000 in 2023. These figures suggest a total decline of 26.6 per cent since 2011, or a decline of around 10,000 horse-owning households per year. The 2023 survey reinforces a sense of decline in horse ownership and the numbers of horses in the UK. Decline has been steady since 2011 and is likely to be a result of the general economic malaise and squeeze on disposable incomes. The British Horse Society also reported in 2023 a 15 per cent drop in the number of rising schools since 2018 from approximately 1,750 to just under 1,500. This was estimated to mean 1.5 million fewer riding lessons taking place in 2023 than 5 years earlier.[35]

Although not prominent among the data used within the horse industry, the UK government does collect and report its own estimate for total horse numbers in the UK. The exercise is as part of international obligations under the Framework Convention on Climate Change where signatory governments of the richest countries (Annex I countries) are required to report annually on a range of sources of emissions of greenhouse gases, including from different types of farm animals ranging from poultry and pigs to beef and dairy cattle, including horses. The Department for Energy Security and Net Zero, formerly the Department for Business, Energy and Industrial Strategy, is responsible for compiling the Annual Inventory and bases its figures on horse numbers on agricultural data from Defra and the BETA surveys. The figures are broken down by horses kept on agricultural holdings, professional horses and domestic horses and are shown in Table 6.2. These are the UK government's official estimates of the total UK horse population and show a marked increase in total numbers, an almost doubling, between 1990 and 2005, from 570,000 horses to 1.036 million. From 2005 to 2021 the figures show a decline

Table 6.2 The UK Government's Estimates of the UK Horse Population, 1990–2021

	1990	*1995*	*2000*	*2005*	*2010*	*2015*	*2016*	*2017*	*2018*	*2019*	*2020*	*2021*
Horse kept on agricultural holdings	202	273	287	346	312	283	268	258	250	251	236	231
Professional horses	62	62	70	91	91	87	87	87	87	87	87	87
Domestic horses	305	348	649	599	621	608	608	608	608	608	608	608
Total	570	684	1006	1036	1024	978	963	954	945	947	932	927

Source: UK Government (2023) *UK Greenhouse Gas Inventory, 1990–2021*. London: Department for Energy Security and Net Zero. Annexes, p.808. https://unfccc.int/documents/627789

of 109,000 or 10 per cent to 927,000.[36] (The most recent BETA figures for 2023 are not yet included.)

The number of regular riders (that is those who ride at least once a month) reported by the BETA surveys has oscillated over recent decades. It rose from 1.4 million in 1999 to 2.1 million in 2005, then fell to 1.6 million in 2011 and 1.3 million in 2015. However, it was estimated to have grown to 1.8 million in 2019 and 1.82 million in 2023. It is estimated that 3.2 million people in 2023 had ridden at least once in the previous 12 months, representing an 18 per cent growth since the 2015 BETA survey, but a 25 per cent decline since the peak of 4.3 million recorded in the 2006 survey.[37] Just over half of all riders are in the south east, south west, East Anglia and Wales. The vast majority of riders ride all year round and for pleasure. Around a third of riders participate in affiliated competitions and almost two-thirds participate in non-affiliated competitions. Around 6 million households contain an ex-horse rider, and ex-riders commonly cite financial reasons for ceasing horse-riding, suggesting participation in horse-riding is influenced by general levels of disposable income and the cost of living.

There were approximately 630,000 privately owned horses in Britain as estimated in the 2023 BETA survey. This remains higher than in 1999 but does represent a 23 per cent decline on 2015 levels. There are an estimated 331,000 horse-owning households, meaning that horse-owning households own just under two horses each on average. The number of horse-owning households has fallen by more than a half since 2006. Horses are owned for an average of 8 years, although over a quarter of horses have been owned for 10 years or more. The average reported purchase price of a horse rose from £1,890 in 2015 to £2,300 in 2019 and £2,800 in 2023.[38] Leisure riding and hacking are the main reasons horses are kept. There has been a general trend away from keeping horses on owners' own land, which has fallen from 50 per cent of owned horses in 2006 to 38 per cent in 2023. The proportion kept in 'do-it-yourself' liveries, where the owner is responsible for their care, has fluctuated from 22 per cent in 2006 to 28 per cent in 2019 and down to 24 per cent in 2023.[39]

Total annual direct expenditure on horses by owners was estimated at £3.4 billion in 2019, or £4,800 per horse and had risen to £3.8 billion or £6,000 per horse in 2023. A quarter of direct expenditure goes on accommodation, with a further fifth on feeds. Approximately 10 per cent goes on healthcare and a further 10 per cent on farriers. Increasing amounts have been spent on paid help to support the care of horses, which more than trebled from £29.9 million in 2015 to £96 million in 2019 but dropped back to £65 million in 2023.[40] There is additional indirect spending which is not essential to the care and upkeep of horses. Spending on clothing, hats and footwear and other equipment, along with spending on books, magazines and digital subscriptions among regular riders, had increased by almost a third since 2015 to an estimated £742 million in 2019, in part reflecting the increasing number of people riding. However, this sort of spending has declined during the post-Covid economic squeeze to an estimated £641 million in 2023.[41] Similarly, spending on riding lessons is estimated to have fallen from £247 million in 2019 to £195 million in 2023.[42]

Problems associated with Brexit, the Covid pandemic and the pressures on disposable incomes from the cost-of-living squeeze have increased economic pressures on the horse sector. Brexit has been a preoccupation of the twice-yearly meetings of the British Horse Council since 2018 as the nature of the UK's withdrawal came into sharper focus and the implications for the costs and administration of horse movement became clearer. Prior to Brexit, the movement of horses across European borders had become a standard practice in horse sports and breeding. The end of the transition arrangement and the UK's exit from the EU has meant that movement of horses between Great Britain and the EU or Northern Ireland is now subject to EU regulatory controls. These require that the owners pre-notify the border authority before the horse is moved and ensure the animal meets pre-export isolation and residency requirements, has an export health certificate and moves through a designated border control post that is able to inspect equines. The House of Commons Environment, Food and Rural Affairs Committee reported in September 2021 on its investigation into post-Brexit animal movements that since January 2021, horse movements for racing and breeding purposes had reduced by 40 to 60 per cent and that the process of moving horses was now considerably more administratively complex and time-consuming.[43] It is also much more expensive. Analysis by the British Equestrian Foundation found that the costs of transporting one horse in a horsebox by Eurotunnel had more than trebled from £1,495 before Brexit to £4,842 after Brexit, while by ferry the costs had more than quadrupled from £995 to £4,157.[44]

The Covid pandemic also impacted the horse economy. Lockdowns curbed the activities of many equestrian businesses, although horses still needed to be cared for. Over 40 per cent of equestrian businesses surveyed in April 2020 reported that they had no income at all, and only 11 per cent reported that the pandemic had not impacted them. Most affected were riding schools, who were most likely to have reduced employees' working hours and made use of the furlough scheme.[45] Lockdown meant there was no racing at Britain's 59 racecourses. The British Horseracing Authority estimated that racecourses lost over £8 million of income per month during the Covid lockdown. Because over 45 per cent of their revenues are generated from race-day attendees, even when racing was able to resume, losses continued. Racehorse trainers reported reduced numbers of horses in training. Total lost revenues for racecourses, trainers, breeders and jockeys were estimated to be over £100 million.[46] The Horseracing Levy Board was able to make £28 million in grants and loans available to racecourses in need of assistance, but Covid left a longer-lasting impact. UK thoroughbred sales were estimated to be down about 50 per cent and fewer foals have been born since.[47]

Horsey Places: Newmarket as a Global Horse Economy Hub

Although the British horse economy is extensive and distributed across the territory of the UK, it has a distinctive geographical pattern. The horse economy is largely part of the rural economy. This means horse businesses operate within local rural economies that are characterised by the twin challenges of sparsity and remoteness. Rural areas have lower population densities than urban areas, and businesses are more likely to be geographically more distant from their clients and suppliers. Rural economies are also distinctive in having larger proportions of businesses that are microbusinesses employing ten or fewer staff. They have higher levels of employment and homeworking. Even within rural regions, there will be a patterning of horse-related businesses, with riding schools, for example, more likely to be located closer to centres of population. Efforts to map the geographical distribution of horses and horse owners show distinct patterns. Horse owners are more densely clustered in London and the south east than are horses, but horses show high geographical concentrations in Sussex, Hampshire, the M4 corridor, Worcestershire, Cheshire and the north west.[48] Then there is the stand-out cluster of Newmarket, the archetypical horsey place in Britain, known within the racing industry as 'the headquarters'.

Newmarket is a small town of 17,000 inhabitants in Suffolk in East Anglia. It is almost an exclave, a curiosity of political geography, being almost surrounded by the county of Cambridgeshire, yet part of the county of Suffolk. It is located 12 miles east of Cambridge within relatively flat former heathland ('the Heath') which is good for racehorse training and horseracing. While previously a small village, it was visited by King John in the early seventeenth century and chosen as the location for the development of a royal palace and as his holiday resort. Under the reign of James I (1603–1625), Newmarket became favoured as Royal Newmarket, and King Charles II (who reigned from 1660 to 1685) was so keen on the racing there that he famously moved his court to Newmarket for the autumn racing season. Newmarket became the location for royal stables and racehorse breeding and went on to become the centre of the British horseracing industry. The Jockey Club was established in 1750 initially as a gentlemen's club, but it soon became the de facto governing body for the sport of racing, "the upholder of turf morality and the distributor of discipline".[49] It leased land in Newmarket in 1752 and went on to establish itself on the High Street. Elite racing institutions, breeders and trainers began to concentrate around Newmarket. Breeders cluster there because of the infrastructure, specialist services and the availability of a specialist trained workforce.[50]

By the early twenty-first century, more than a third of the town's adult population was involved directly in the racing industry.[51] It is a town "unmistakably concerned with a single industry".[52] The town has two racecourses which support more horseracing events than anywhere else in the UK. The two racecourses typically attract around 350,000 attendees at racing events each year and have a total turnover of around £20 million. The National Horseracing Museum was established in the 1980s and hosts a national collection of pieces on the history and science of racehorses. The National Stud is owned by the Jockey Club and attracts around 20,000 visitors each year and serves as a visitor attraction and a training centre as well as a stud. Tattersalls, the main racehorse auctioneers for the UK and Ireland, are also based in Newmarket and have total sales of some £300 million a year. Newmarket has 70 trainers located around the town, and the Newmarket Stud Farmers' Association has almost 60 stud farms as members. The cluster is home to 3,000 horses, many of which are among the world's most elite racehorses. It is the dominant force in UK racing, with Newmarket-trained horses winning almost three-quarters of the top competitive races on average.[53]

Anthropologist, Rebecca Cassidy, has carried out detailed ethnographic work to understand culture and practice in the thoroughbred world of Newmarket.[54] She describes Newmarket as a town in which "everything is horse".[55] The horse world is a relatively exclusive one with a strong self-image, where the ability to 'talk horse' is crucial to access and acceptance. Attachment to racing means everyone has something in common inside horse society. The industry dominates the town and its surroundings, with the Jockey Club headquarters and the National Horseracing Museum at its centre. There are 57 miles of horse-walks that criss-cross the town with a traffic light system at road crossings that "announce the horse's primacy in the town".[56] Drivers through the town can expect to encounter racehorses coming and going, especially in the mornings. Artificial gallops are separated by white railings, and the morning landscape can be dotted with hundreds of horses and riders. The town sits in the middle of a network of stud farms and racing stables in the surrounding countryside. These all require land for grazing and exercising. The surrounding area is characterised by "striped green expanses, clipped hedgerows, painted fences and the sweeping driveways of the stud farm", a highly manicured landscape that represents "some of the neatest countryside imaginable",[57] what Jonathan Meades has described as "clipped immaculacy" and "fanatical tidiness".[58] Horseracing enjoyed a boom of investment from Middle Eastern investors in the 1970s, and elite horse-breeding is at the centre of racing and Newmarket's economic system. To the casual visitor, the town has an affluent feel, and it has been remarked that Newmarket is the smallest population in Britain to sustain a Waitrose supermarket.[59] Although racing success is important for prestige, money from breeding usually outweighs prize money and most racehorse owners lose money on their investments.

England's most prestigious studs are found around Newmarket, and approximately 9,000 acres of manicured land make up the surrounding 'studscape'.[60] Several global thoroughbred breeding companies are based around the town. Sheik Mohammed bin Rashid Al Maktoum, the vice president and prime minister of the United Arab Emirates and the crowned ruler of Dubai, owns several studs in England, including in Newmarket. Other influential Newmarket breeders have included Saudi Arabia's Prince Khalid Abdullah (cousin of the late King Fahd). The small East Anglian town has therefore become the centre of a global elite in the world of thoroughbred breeding and racing. Meades described the town as "weirdly international" and having more in common with the likes of Klosters or Monte Carlo than its neighbouring towns of Thetford or Bury St Edmunds.[61]

Cassidy's ethnographic work portrays a relatively conservative and insular economic community in Newmarket centred on horse-racing, but with a range of social classes involved. At one extreme are some of the wealthiest individuals in the world who own studs. At the other extreme are the many grooms and stable hands who work long hours, often for low pay, but make up a vibrant and hedonistic occupational community frequenting the town's high street pubs. "The camaraderie of the yards of Newmarket can be thrilling, and this goes some way toward accounting for the fulfilment and excitement that such a lifestyle can sometimes offer, despite its lack of obvious rewards".[62] Trainers, on the other hand, rarely frequent the town centre pubs, instead of enjoying the gastropubs of Newmarket's surrounding villages. Racing society, the owners, breeders and trainers, is generally wealthy and traditionalist. The area elects Conservative MPs, and there has long been a Eurosceptic vein of political feeling. In the 1990s, the area was marked by relatively strong support for the Referendum Party, a single-issue, Eurosceptic party which made the case for a referendum on the UK's relationship with the European Union.[63] In the 1997 General Election, the Referendum Party received more than double its national average of the vote in Newmarket.[64] In the 2016 Brexit Referendum, the area had one of the highest majorities for voting 'Leave'.[65] Cassidy wrote:

> Newmarket has allowed a particular way of life to be preserved. The High Street and its abandonment to the lads, the preservation of the yards, the Health, and the stud land reflect the same kinds of relationships that gave rise to this landscape. The status of Newmarket as "Headquarters" depends upon the continuing relevance of these connections with the past.[66]

A study by economic development consultants in 2014 estimated that the horse-racing industry contributed some £200 million a year to the local economy of the area and supported 3,285 jobs directly and approximately 8,500 in total. The largest employers were horseracing trainers who employed over 2,000 full-time equivalents, followed by stud farms, horseracing institutions and scientific and veterinary institutions. Overall, the sector represented some 6 per cent of the total gross value added of the two closest local authority districts although it has been estimated that a much higher proportion of local employment is dependent upon the horse industry.[67] The locality is a rare example of a local economy in a rural region developing a specialist sectoral cluster. Its competitive edge is based on its locally rooted, highly specialist and distinctive resources, the skills and infrastructure for horse-breeding, training and racing. These endogenous economic resources help bring in trade, investment and visitors from elsewhere. Money flows into Newmarket from around the national community of horseracing spectators and the horse economy. It is also part of a global network of elite horseracing businesses, with thoroughbred racehorses transported in and out of Newmarket around the world.[68] It is a model of what has been called 'neoendogenous rural development' that can also be identified in the business clusters based on second-hand books in Hay-on-Wye or gastro-dining in places like Ludlow.[69] It is, however, a distinctive local economy that has been over four centuries in the making and so not a model that can easily be replicated elsewhere.

Managing the Horse Economy

As we noted in Chapter 5, the horse economy has evolved over the past half century in the context of a countryside that has shifted from a place of production to one of consumption. Horse enterprises have been part of an economic development strategy supported by policies that encourage economic diversification away from a reliance on primary industries such as agriculture. Horse-based leisure and tourism are common features of rural diversification and economic development across rural Europe. Horse-riding holidays may involve 'based rides' where holidaymakers return to a central base each day or 'trail rides' where visitors ride to a different venue each day. In 1975, the International Federation for Equestrian Tourism was created to provide an international network to support the development of equestrian tourism and promote best practice. It has promoted the coordination of transnational equestrian trails and routes and worked to support collaboration between small tourism businesses to strategically develop visitor experiences.[70] In the UK, 62 per cent of adults have taken a holiday to specifically pursue a hobby or interest at some point, and 17 per cent in the 5-year period up to 2019. More than one in eight (13 per cent) of those had taken an adventure holiday in the past 5 years. Of those people, 12 per cent had taken horse-based holidays such as riding or pony trekking.[71]

Competing demands upon rural land, coupled with rising environmental concerns, led to a sense from the 1980s of the countryside and rural economy being at a crossroads. Policy-makers struggled with how competing pressures were to be reconciled and how a new rural economy might be managed and supported to help develop local livelihoods in the face of marked structural change. From the 1940s to the 1990s, there had been no Rural White Papers, only agricultural ones, but John Major's Conservative Government produced a Rural White Paper in 1995 and New Labour followed with a vision of their own in 2000. Each of these tried to grapple with competing social demands upon rural space. New Labour had been elected on a modernising mission, although the party had not thought much about its approach to the countryside and rural land use prior to its election. The New Labour Government sought to develop a modernising vision for rural areas to help dispel the argument that it was ignorant about, or hostile to, rural interests. It was faced with a farm income crisis and several calls for financial bailouts from the farming unions in its first two years in office. It settled on a progressive rural development agenda that would work with the grain of social and economic change and support a vibrant and diversifying rural economy around which farming, environmental and economic development interests might be able to align.[72]

One Labour minister appointed to the Ministry of Agriculture, Fisheries and Food, Lord Donoughue, was a former academic and head of the Number 10 Policy Unit in the Labour Government of the 1970s. A horse-racing enthusiast and racehorse owner, he took an active interest in the horse industry and how it was represented and governed.[73] New Labour wanted to broaden rural policy beyond the traditional constituencies of the agricultural industry and the environmental lobby. It more closely involved consumer and environmental groups in rural policy development, and Donoughue was also interested in developing a more unified approach to the horse industry.[74] He pressed horse organisations to create an umbrella body to represent the sector as a whole.[75] In 1999 the British Horse Industry Confederation was launched, bringing together the British Horseracing Board, the British Equestrian Federation and the British Thoroughbred Breeders' Association. In the debates around the 2000 Rural White Paper, the British horse industry presented itself as the second largest industry after farming in rural areas. Evidence was given collectively to a Parliamentary Select Committee looking into the Rural White Paper by the new British Horse Industry Confederation. They pressed government to designate a minister as having lead responsibility for the horse industry. There were concerns about the difficulties in securing designation for bridleways, the economic challenges facing the then 2,000 or so riding schools and 7,000 horse-breeders. It was also emphasised that betting on horses raised about £450 million a year in tax and betting duty at this time.[76] It was notable that for the first time in a major piece of national rural policy, the horse industry spoke authoritatively and seemingly with one voice.

Following the publication of New Labour's Rural White Paper in November 2000, rural affairs and land use policy was swept up in the Foot and Mouth Disease (FMD) crisis which broke out in spring 2001 and dominated countryside politics for the rest of the year. In the aftermath of FMD there was renewed interest in overhauling the government's approach to agriculture and the rural economy. The Government created a role of 'Minister for the Horse', and in 2002 it established a Horse Industry Team within the civil service to help co-ordinate policy on areas affecting the horse industry. In 2003, the new government department, Defra, the Department for Environment, Food and Rural Affairs, commissioned a study of the horse industry in order to underpin a long-term strategy.[77] The project sought to develop a workable definition of the horse industry, provide a baseline assessment of the industry's economic size and highlight the key strategic issues for the development of the industry over the following 10 years. It defined the horse industry as "encompassing all activity that has the horse as its focus and activity that, in some reasonable capacity, caters for such an industry".[78]

Coming to a judgement on the economic size of the industry at that time was difficult, not least because of the potential extent of the informal economy around horses. The study produced some headline estimates of the size of the industry at this point in time. It estimated that there were around 50,000 people directly employed in the horse industry and when indirect employment (i.e. jobs supported by expenditure in the horse industry) was included, then total employment was likely to be of the order of 150,000–250,000 jobs. Five million people, or 11 per cent of the population, were estimated to have an active interest in horses, which excluded those who only watched racing on TV. Almost half of those, some 2.4 million people, rode horses. It considered five other recent estimates of the economic scale of the horse industry and concluded with its own best estimate that the industry represented approximately £3.4 billion of gross value added.[79] The gross output of the horse industry was estimated to be over a quarter of that of the agriculture industry.[80] The study placed the British horse industry in an international comparative context. With a horse population of 1.0 to 1.7 horses per 100 people, there were proportionally fewer horses in Britain than in the United States (2.4 horses per 100 people) or Australia (estimated at between 4.6 and 9.2).[81]

The study was able to estimate the numbers of various types of consumers in the horse industry. There were an estimated 500,000 domestic horse owners at this time, and of the 2.4 million riders around half were estimated to be consumers of riding lessons. Some 375,000 riders were estimated to hire horses for trekking or holidays. An estimated 600,000 riders participated in equestrian sporting events, and there were 900,000 readers of horse magazines.[82] Data from a large national consumer survey suggested that 10.7 million people had some interest in horses, which might include watching racing on TV, representing 23 per cent of the adult population. The regional breakdown suggested the proportions were highest in East Anglia (26 per cent), the south west (24 per cent), north west (24 per cent) and Yorkshire (24 per cent) and lowest in the West Midlands (21 per cent).[83] The survey suggested that the level of interest in horses in the early years of the twenty-first century was growing (up from 20.4 per cent in 1998 to 22.6 per cent in 2003). The study estimated that there were between 600,000 and 975,000 horses in Great Britain. (It also estimated that there were around 29,000 in Northern Ireland, bringing the upper estimate to a million.)[84] Some 30,000 horses were owned by riding stables, livery yards and trekking centres; around 100,000 were involved in sporting competitions, while it was estimated that there were around 40,000 racehorses in the country.[85]

Following the baseline work, Defra, with the British Horse Industry Confederation, published a Strategy for the Horse Industry for England and Wales.[86] Its aim was "to foster a robust and sustainable horse industry, increase its economic value, enhance the welfare of the horse, and develop the industry's contribution to the cultural, social, educational, health and sporting life of the nation".[87] The Strategy celebrated the fact that the relationship between government and the horse industry had developed significantly over the preceding four years and set out a vision for the next decade. It sought to widen participation in horse-related activities, promote the growth and development of the industry, help develop sporting success and strengthen British competitiveness in sporting and business terms. The Strategy was rooted in a concern that the industry had previously suffered from being fragmented and lacked overall strategic direction. Its first aim was therefore to encourage the main horse industry representative bodies to better coordinate with each other. It called for better research into riders and consumer trends and stronger coordination of the provision of services by riding schools. It estimated that there were around 18,000 equestrian businesses in Britain, which were mainly small businesses and were generally less profitable than other sectors. There was, therefore, a challenge to improve the business performance of the sector and enhance the value of services provided.

The Strategy included an aim of encouraging sporting excellence. Sporting excellence requires clear plans, good marketing, development programmes for horses and riders, good facilities, talent-spotting, coaching and advanced sports science support. The national strategy emphasised extending coaching development, improving facilities and extending long-term athlete and equine development programmes. It also sought to encourage unaffiliated bodies to engage with the main national governing bodies.[88]

Although the period 1997 to 2002 had seen a significant reorganisation in how the horse industry was represented, a key divide remained between horse-racing and non-racing sub-sectors. Funds raised from betting have tended to remain within the racing sector, while representative bodies are supported by membership subscriptions. Horse-breeding remains highly fragmented, with a large number of breeders having just a few breeding mares. Evidence suggested the horse industry was growing in terms of turnover and the number of people participating in equestrian activities was growing. In 2006, the British Horse Industry Confederation were bullish about growth and prospects, reporting that the horse industry was "growing at significant pace".[89] The period since 2010 has been more challenging, however.

Although Covid left its mark on the horse economy, the subsequent economic squeeze in the aftermath of Brexit, Covid and the effects of the Ukraine war on the economic climate has intensified pressures. High inflation, low economic growth and rising interest rates have brought pressure on households and businesses, with fuel costs in particular increasing markedly. For those who own horses, costs have to be born in the short term, but for those whose spending in the horse economy is more discretionary, such as the clients of riding schools, general economic belt-tightening has meant reduced horse business incomes. Several governing bodies of horse sports announced measures in 2022 to provide financial support as concerns grew about low levels of entries to equestrian sporting competitions.[90] Horse welfare organisations are also concerned that the economic squeeze will mean more unwanted horses having to be accommodated.

By December 2022, cost pressures were becoming acute. Nicole Cooper, the General Manager of the Carrington Riding Centre near Manchester, explained to the BBC how the gas and electricity bill for her business had doubled from £1,500 a month to £3,000. Haylage had more than doubled in price from £30 a bale to £65. The cost of wood shavings had more than quadrupled from £2.50 to £10.20. The cost of looking after her own horse had risen from £250 per month in 2013 to £650.[91] The implications of the economic squeeze for the horse economy became a running theme in the news pages of *Horse & Hound*, the largest circulation publication of horse riders and owners, during 2022, including the declining number of rising schools, and there was a sense of mounting economic pressures. The Chief Executive of World Horse Welfare, Roly Owers, warned "It is difficult not to be hugely concerned about the rising cost of living". The Executive Director of BETA, Claire Williams, said: "It's a very challenging time, but the industry long-term does tend to be resilient … At the end of it, we all care for our horses".[92] By the end of 2022, there was some evidence that horse owners were prioritising expenditure on their horses and cutting back elsewhere to cope with the sharp increase in living costs and general inflation.[93]

Conclusions

As we saw in Chapter 5, total horse numbers in Britain declined during the twentieth century, initially due to the growth of the motor vehicle in urban public transport but also because of the gradual adoption of the tractor in agriculture. The horse economy was therefore steadily shrinking in size from the early twentieth century through to the 1970s. At the end of the Second World War, there were still over 500,000 horses being used in British agriculture,[94] but these numbers reduced markedly in the post-war decades. The twentieth century saw the radical transformation of the horse in British economic life, from a creature of agriculture, industry and transport to an animal associated primarily with sport and leisure. Although this shift has been associated with a decline in total horse numbers, the horse sector is still economically significant, ranging from a globally integrated elite horseracing sector at one extreme, to extensive participation in local horse-riding for pleasure at the other. A global industry of some £250 billion annual turnover, with 7 million horses in Europe and millions of hectares of land, means the horse economy still has some impact and influence upon land use and can be a novel and distinctive form of low-impact rural tourism. In some key locations such as Newmarket in the UK, the Inner Bluegrass region of Kentucky in the United States or the Upper Hunter region of New South Wales in Australia, the horse economy can be a prominent force in local economic development.[95]

It is difficult to be clear on total horse numbers in Britain, but the various national surveys and estimates that have been produced suggest that the total number of horses during the first two decades of the twenty-first century is significantly higher than that in the 1980s and 1990s. The UK government estimates an almost doubling of total horse numbers between 1990 and 2005, and this growth was driven by general economic growth and the development of a stronger recreation and leisure component of the rural economy. From the late-1990s, BETA's surveys of the horse industry helped to introduce some standardisation into data collection. From a high point over the first decade of the twenty-first century, total numbers have fallen back. National economic growth has been weak under conditions of austerity since 2010, and the rising cost of living and aftermath of Brexit and Covid have all contributed to a squeeze. Riding Schools are reported as struggling and significant numbers have closed since 2018.

The horse economy was actively promoted for a period at the turn of the twenty-first century both as a strategy for farm diversification and the development of a rural economy more oriented to recreation and leisure industries. However, there has been less strategic interest in the support and development of the sector since 2010, when governments have been more concerned with letting market forces determine the complexion of rural economies. While leisure riding remains popular, changes in gambling practices and markets bring economic pressures on the horseracing industry. Brexit, Covid and the general economic squeeze in living standards all contribute to harder times for the UK horse economy.

Notes

1 https://www.equinebusinessassociation.com/equine-industry-statistics/
2 Equine Business Association. https://www.equinebusinessassociation.com/equine-industry-statistics/
3 World Horse Welfare and EuroGroup4Animals (2015) *Removing the Blinkers: The Health and welfare of European Equidae in 2015*. Snetteron: World Horse Network.
4 JDA Research (2023) *The National Equestrian Survey 2023. Overview Report.* Wakefield: British Equestrian Trade Association.
5 The figure is quoted in Crossman, G. (2010) *The Organisational Landscape of the English Horse Industry: A Contrast with Sweden and the Netherlands*. University of Exeter PhD thesis, p.17.
6 British Horse Council (2019) *The British Horse Sector – Why it matters for the 2019 General Election*. High Wycombe: British Horse Council.
7 Bell, S. (2020) Horse racing in imperial Rome: Athletic competition, equine performance, and urban spectacle, *International Journal of the History of Sport* 37, 183–232.
8 Huggins, M. (2018) *Horse Racing and British Society in the Long Eighteenth Century*. Woodbridge: Boydell Press.
9 PWC (2018) *The Contribution of Thoroughbred Breeding to the UK Economy and Factors Impacting the Industry's Supply Chain*. Newmarket: Thoroughbred Breeders Association.
10 https://www.internationalstudbook.com/about-isbc/
11 Pickerill, J. (2005) '95% of thoroughbreds linked to one superstud', *New Scientist* 6 September, https://www.newscientist.com/article/dn7941-95-of-thoroughbreds-linked-to-one-superstud, quoted in McManus, P. *et al.* (2013) *The Global Horseracing Industry: Social, Economic, Environmental and Ethical Perspectives*. London: Routledge, p.16.
12 Artificial insemination is prohibited in order to manage the breeding process. There are two mating seasons – one for the northern hemisphere and one for the southern. Breeding stallions are transported around the world and can make with hundreds of mares per season. There are concerns about the level of inbreeding among thoroughbred racehorses – see McGivney, B. *et al.* (2020) Genomic inbreeding trends, influential sire lines and selection in the global Thoroughbred horse population, *Nature Scientific Reports* 10, 466.
13 McManus *et al.* (2013); Cassidy, R. (2007) *Horse People: Thoroughbred Culture in Lexington and Newmarket*. Baltimore: Johns Hopkins University Press.

14 Thoroughbred Breeders' Association (2023) Third thoroughbred breeding industry economic impact study provides blueprint for future progress, *Press Release* 18th January.
15 PWC (2018), p.8.
16 PWC (2018), p.8.
17 Deloitte (2013) *Economic Impact of British Horseracing 2013*. London: British Horseracing Authority; PwC (2018).
18 A guinea is worth £1.05. It is a tradition in horseracing to price horses in guineas rather than pounds.
19 British Horseracing Authority (2020) Written Evidence to the House of Commons Culture, Media and Sport Committee Inquiry into *The Impact of Covid-19 on DCMS Sectors*. Session 2019–21. London: Stationary Office.
20 Cassidy, R. (2020) *Vicious Games: Capitalism and Gambling*. London: Pluto, pp.68–9.
21 British Horseracing Authority (2020).
22 PWC (2018), p.7.
23 PWC (2018), p.9.
24 PWC (2018), p.9.
25 Cassidy (2020), p.67.
26 Frontier Economics (2016) *An Economic Analysis of the Funding of Horseracing*. Report to the Department for Culture, Media and Sport. London: Frontier Economics.
27 Morgan Stanley (2015) *Global Gaming Report 2015*, quoted in Cassidy (2020), p.7.
28 Walmsley (2012); Walmsley, D. (2021) *Mintel Gambling Review UK 2021*. London: Mintel.
29 Cassidy (2020), p.78.
30 Cassidy (2020), p.136.
31 Frontier Economics (2016), p.7; PWC (2018), p.9; Horserace Betting Levy Board (2023) *Annual Report and Accounts 2021/22*. London: HBLB.
32 British Horseracing Authority (2020).
33 Crossman, G. and Walsh, R. (2011) The changing role of the horse: From beast of burden to partner in sport and recreation, *International Journal of Sport and Society* 2, 95–110. Membership figures are organisations' websites (accessed 6th July 2023).
34 JDA Research (2019) *The National Equestrian Survey 2019. Overview Report*. Wakefield: British Equestrian Trade Association; JDA Research (2023).
35 Jones, E. (2023) Riding at risk as schools close but participation slightly up, *Horse & Hound*, 9 March p.4.
36 UK Government (2023) *UK Greenhouse Gas Inventory, 1990–2021*. London: Department for Energy Security and Net Zero. Annexes, p.808. https://unfccc.int/documents/627789
37 JDA Research (2019), p.9.
38 JDA Research (2023), p.30.
39 JDA Research (2019), pp.22–28; JDA Research (2023), p.32.
40 JDA Research (2019), pp.32–33; JDA Research (2023), p.39.
41 JDA Research (2023), p.40.
42 JDA Research (2023), p.41.
43 House of Commons Environment, Food and Rural Affairs Committee (2021) *Moving Animals Across Borders*. HC 79. London: Stationary Office, p.17.
44 British Equestrian Federation (2020) Costs of travelling horses to Europe post-Brexit, *News Release* https://www.britishequestrian.org.uk/news
45 British Grooms Association and Equestrian Employers Association (2020) *Coronavirus Impact Survey*.
46 British Horseracing Authority (2020).
47 Davies, E. *et al.* (2020) The impact of Covid-19 on staff working practices in UK horseracing, *Animals* 10(11), 2003.

48 Boden, L. *et al.* (2012) Summary of current knowledge of the size and spatial distribution of the horse population within Great Britain, *BMC Veterinary Research* 8, 43.
49 Cassidy (2007), p.57.
50 Cassidy (2007), p.53.
51 Cassidy (2007), p.ix.
52 Cassidy (2007), p.64.
53 Cassidy (2007), p.63.
54 Cassidy, R. (2002) *The Sport of Kings: Kinship, Class and Thoroughbred Breeding in Newmarket*. Cambridge: Cambridge University Press.
55 Cassidy (2002), p.13.
56 Meads, J. (1997) *Even Further Abroad – Nag, Nag, Nag*, BBC Two
57 Cassidy (2002), pp.19–20.
58 Meads (1997).
59 Cassidy (2002), p.28.
60 Cassidy (2007), p.60.
61 Meads (1997).
62 Cassidy (2007), p.62.
63 Ford, R. and Goodwin, M. (2014) *Revolt on the Right: Explaining Support for the Radical Right in Britain*. London: Routledge.
64 Cassidy (2002), p.46.
65 Forest Health, the district which includes Newmarket, voted Leave with 65 per cent of the vote, ranking it 41st out of 382 local authority areas in terms of the strength of the Leave vote.
66 Cassidy (2007), p.64.
67 SQW (2014) *Newmarket's Equine Cluster: The Economic Impact of the Horseracing Industry Centred upon Newmarket*. Cambridge: SQW.
68 McManus, P. *et al.* (2013) *The Global Horseracing Industry: Social, Economic, Environmental and Ethical Perspectives*. London: Routledge.
69 Lowe, P. *et al.* (1995) Beyond endogenous and exogenous models: Networks in rural development, pp.87–105 in J.D. van der Ploeg and G. van Dijk (eds.) *Beyond Modernization: The Impact of Endogenous Rural Development*, Assen, Netherlands: Van Gorcum.
70 European Horse Network (2017) *Equestrian Tourism*. Briefing Note, July 2017. Brussels: European Horse Network https://www.europeanhorsenetwork.eu/documents/position-statements/
71 Walmsley, D. (2019) *Special Interest Holidays UK*. London: Mintel.
72 Ward, N. and Lowe, P. (2007) Blairite modernisation and countryside policy, *The Political Quarterly* 78 (3), 412–21.
73 Donoughue, B. (2016) *Westminster Diary Volume 1: A Reluctant Minister Under Tony Blair*. London: I.B. Tauris; Donoughue, B. (2018) *Westminster Diary Volume 2: Farewell to Office*. London: I.B. Tauris.
74 Donoughue (2016), p.241.
75 Donoughue (2018), p.125.
76 Clayton, M. *et al.* (1999) Oral evidence, House of Commons Select Committee on Environment, Transport and Regional Affairs *The Rural White Paper*, Session 1999–2000, HC Paper 32, London: The Stationery Office.
77 The Henley Centre (2004) *A Report of Research on the Horse Industry in Great Britain*. Report to Defra and the British Horse Industry Confederation. London: Defra.
78 The Henley Centre (2004), p.1.
79 The Henley Centre (2004), p.3.
80 The Henley Centre (2004), p.35.
81 The Henley Centre (2004), p.34.
82 The Henley Centre (2004), p.40.

83 The Henley Centre (2004), p.42.
84 The Henley Centre (2004), p.34 & p.45.
85 The Henley Centre (2004), p.45.
86 British Horse Industry Confederation, Defra, Department for Culture, Media and Sport and the Welsh Assembly Government (2005) *Strategy for the Horse Industry in England and Wales*. London: Defra.
87 British Horse Industry Confederation *et al.* (2005), p.18.
88 British Horse Industry Confederation *et al.* (2005), pp.85–88.
89 British Horse Industry Confederation (2008) Memorandum of evidence, pp.169–71 in House of Common Environment, Food and Rural Affairs Committee (2008) *The Potential of England's Rural Economy*. HC544-II. Session 2007/08. London: The Stationary Office.
90 Murray, R. (2022) Support introduced to help sports cope with rising costs, *Horse & Hound* 16 June, p.6.
91 BBC Radio 4, *Farming Today*, Monday 28th November 2022.
92 Elder, L. (2022) How the economy is affecting the horse world, *Horse & Hound* 29 September, pp.4–5.
93 Jones, E. (2022) Owners put their horses first as economic situation bites, *Horse & Hound* 22 December, p.10.
94 Murray, K. (1955) *Agriculture - History of the Second World War Series*. London: HMSO, p.274.
95 McManus *et al.* (2013).

7 Horses and Social Change

Introduction

In the early twentieth century, there were 3.3 million horses in the UK. Eight decades later, in 1990, that total had dropped by more than 80 per cent to around 570,000, then almost doubling to just over a million by 2005. The twentieth century saw a dramatic change in the profile of horses in day-to-day life in towns and cities and in the countryside. By the 1950s, most working horses left were in agriculture, but these petered out and became outnumbered by horses used for leisure and sport. The changing place of horses was a function of wider social changes. With the exception of some travelling communities, horse ownership came to be popularly associated with the more affluent middle classes. The transformation to a post-industrial equine world also meant changes in the social relations and institutional frameworks around the horse in Britain. Breeding became more focussed on elite racing and recreational horses, with less emphasis on the heavy working horses to the extent that they became extremely rare. New organisations developed to manage the racing economy and the flow of money that came from regulating betting on horses from the 1960s. The development of other equestrian sports brought new institutions to promote and regulate them, while the growth in horse-riding as a leisure pursuit brought new bodies to promote and support the businesses providing equestrian services to the public.

In the twentieth century, the 'horse world' in Britain continued to enjoy its annual calendar of sporting and social events, some of which have become national institutions. The Cheltenham Festival and the Grand National remained highlights of the horse-racing calendar. The Badminton Horse Trials and Olympic equestrian sport became popular televised sporting occasions. Betting shops emerged as a feature of the high street, and 'a flutter on the horses' became a common feature of late-twentieth-century industrial working-class leisure and popular culture. Horse-riding became more popularised, and participation expanded across social classes. Horse society may have retained its elitist air, with its close associations with royalty and the upper classes, but the numbers of people with some interest in the horse came to be measured in tens of millions. As the institutions involved in horse affairs

DOI: 10.4324/9781003454359-7

became more numerous and specialised, differences between the different parts of the horse world had to be grappled with, which ranged from globally elite thoroughbred racehorse breeding, training and racing through to the trade in horses among travelling communities.

Challenges and controversies flared up and led to change in the governance of the horse, and efforts to bring the sector together into a coherent and unified voice waxed and waned. The sector developed initiatives to widen participation in equestrianism and make the horse world more socially inclusive, while still facing criticism for elitism. The controversy around banning hunting with hounds at the turn of the twenty-first century brought a wave of protest and political activism among equestrian organisations, but also tensions within the horse-riding community about the rights and wrongs of hunting wild mammals with dogs. Most recently, the rise of animal welfare politics has brought challenges for the horse world as the ethics and regulation of equestrian sports including horse-racing have come to be increasingly questioned. The use of horses has also developed in novel ways to support human wellbeing through equine-assisted therapeutic activities, while horses are still used by police forces to manage crowds on protest marches as instruments of public control.

This chapter considers the role of the horse in the context of social change. It examines the institutions of the horse and the politics of managing the equine affairs of the nation. It explores the characteristics of the people who own horses, ride horses and make up 'horse society'. It considers some recent controversies around uses of the horse as well as initiatives to enhance human wellbeing through interaction with horses. In contrast to these emancipatory and wellbeing-enhancing practices, it also considers the continuing role of the horse as an instrument of social control.

The Institutional Geography of Equestrianism

The horse sector in the UK is represented by a complex network of organisations that has evolved over centuries. Although, as we saw in Chapter 6, there have been efforts over recent decades to strengthen the integration and coordination of the sector so that it is able to 'speak with one voice', these have met with only limited success and despite the existence of over-arching bodies, the horse world remains riven by sharp distinctions, especially between the worlds of horseracing on the one hand and those of recreational horse-riding on the other. At one extreme of the spectrum is a commercial industry operating in a highly internationalised and competitive world and involving a global elite. At the other extreme are the users of riding schools and the large number of rural families, farmers and smallholders who might keep a horse or two in a paddock attached to their home and ride just for pleasure. Horseracing enjoys a dedicated income stream in the form of a levy on betting, a rare example of a hypothecated tax, but the wider recreational horse sector receives very little funding from government. In between are the other equestrian sports, the trade association for horse-related businesses, a wide range of horse breeders, the equine veterinary profession and the owners of riding schools and liveries. The spectrum can be thought of as an economic one, but also has a sociological dimension. Whether performed intentionally or not, there remains a sense of elitism that clings to horseracing and a sense of an inferiority complex that is sometimes explicitly cited by those less prestigious, and less well-resourced, parts of the horse world.

The Jockey Club was established in the eighteenth century as an exclusive social club and played the leading role in the governance of horseracing for over 250 years. It now owns 15 racecourses in Britain and the National Stud although its regulatory functions have evolved and been taken over after the establishment of the British Horseracing Board in 1992, now known as the British Horseracing Authority (BHA). The BHA is responsible for the governance, administration and regulation of horseracing in Britain. It normally has an annual income of around £35 million and has almost 300 staff. It is a company limited by guarantee, whose members include the Racecourse Association, the Racehorse Owners' Association, the Thoroughbred Breeders' Association and a licensed personal member who jointly represents the National Trainers Federation, the Professional Jockey Association and the National Association of Racing Staff.[1] The Betting Levy Act 1961 introduced the levy which is managed by the Horserace Betting Levy Board. The money (some £97 million in 2021/22) is used to contribute to prize money but also to support breeding, veterinary science and education and measures to develop horseracing.[2] Although the sums raised by the levy in Britain are relatively modest compared to the sums enjoyed by horseracing in other parts of the world, it does mean that horseracing is relatively well financed in Britain compared to other parts of the horse industry.

Other equestrian sports are represented by British Equestrian, established in 1972 covering dressage, showjumping and eventing (formerly known as horse trials). These equestrian sports have come to be successful Olympic sports in recent decades. Their international competitiveness as Olympic sports mean that they attract national funding through Sport England and its other counterpart bodies. British Equestrian claims to represent the interests of 3 million horse riders, although the member bodies of the various specialist disciplines amount to closer to 200,000 participants.[3] It normally has a turnover of £5–6 million and has around 20 staff.

The horse breeding sub-sector contains a further set of organisations. The National Stud was established after the British government became more directly involved in thoroughbred horse breeding during the First World War. This was driven by a concern about the supply of horses for military purposes, but after the war, the military use of horses sharply declined. The government retained the National Stud, which was moved to Newmarket in the 1960s, until 2008 when its ownership was transferred to the Jockey Club. The Thoroughbred Breeders' Association was founded in 1917 in response to concerns about the challenges posed to horse-breeding during the First World War. There are approximately 3,000 breeders although the vast majority are smaller scale, with five or fewer breeding mares. It is estimated that almost 14,000 hectares of permanent grassland is used for grazing for thoroughbred horses, and the breeding sector supports over 19,000 jobs.[4] Thoroughbred breeding is particularly concentrated in the East Anglian region centred on Suffolk and Norfolk and particularly clustered around Newmarket.

The rest of the recreational horse sector is represented by the British Horse Society (BHS). The BHS was founded in 1947 and is the largest membership equine organisation in the UK with over 119,000 members, although this is still a relatively small proportion of all people who regularly ride horses. Its mission is to promote and advance education, training and safety in all matters relating to the horse, to promote healthy recreation involving the horse, to promote animal welfare and to secure the provision, protection and preservation of rights of way and access for horse-riders. Historically, a key role has been its responsibility for accrediting riding schools. It has an annual turnover of around £12 million and a staff of around 90. Some in the horse sector have viewed the BHS as having too broad a set of interests, and during the late 1990s the organisation went through some upheaval as several of its then component parts became independent, thus fragmenting the representation of the sector. From 2004, its Chief Executive was Graham Cory, the former lead civil servant in Defra responsible for the horse, and this helped improve the BHS's relationship and influence with government departments. In 1992, the National Equine Forum (NEF) was established because the horse industry was widely seen as uncoordinated. It is a fairly modest organisation whose administration was initially funded by the British Association for Riding Schools. It began to hold an annual meeting with a written report of proceedings published as a means of sharing insights and information across the various parts of the sector. It has become customary for the Government Minister responsible for the horse to address the NEF, which celebrated its 30th anniversary in 2022.

As we noted in Chapter 6, there was some reorganisation of the representation of the horse industry in the late 1990s in the early years of the New Labour Government, prompted by the efforts of Lord Donoughue, then minister of farming and the food industry in the first Blair Government. On the one hand, the industry was seen as being fragmented and ineffective in lobbying government. At the same time, responsibilities for the horse were also spread across different government departments including the Home Office, the Department for Culture, Media and Sport and the Ministry of Agriculture, Fisheries and Food (MAFF). In other countries, the horse industry is more firmly located as a subsector of the agricultural industry and the agricultural policy community.[5] Donoughue judged that a precursor to a more strategic and coherent approach from government to the development of the industry was for the industry to demonstrate that it could present a unified voice to government. His efforts in preparing the ground for a strengthened MAFF role as 'Ministry for the Horse' took place against a backdrop of some internal opposition within MAFF.[6] He was, however, able to persuade the key industry bodies of the need for a new unified voice for the sector, and in early 1999 the British Horse Industry Confederation (BHIC) was launched. After the BHIC was established, a team of civil servants was formed to support the strategic development of the sector, and between 1999 and 2006 the horse industry became more prominent within national rural affairs and rural economic development. However, after the Strategy for the Horse was produced in 2005, momentum was lost. There was little follow-through in implementing the measures set out in the Strategy, as well as in monitoring progress. By 2010 and the change of Government, horse industry representatives were already bemoaning the loss of impetus.[7] Following the 2008 financial crisis and the subsequent austerity measures introduced by the Coalition Government, rural affairs including the development of the horse industry were pared back within Defra and never recovered.

In 2018, the BHIC and the Equine Sector Council for Health and Welfare combined to form the British Horse Council which presents itself as 'the voice of the equine sector'. It is a Community Interest Company comprised of seven directors drawn from the main horse industry bodies. It holds two meetings a year which receive updates from across various parts of the horse industry, and mainly focuses on work where there is an interest across the sector such as disease surveillance, horse identification and traceability arrangements, Brexit and Covid. There continue to be difficulties and fragmentation around the horse sector, although the annual meetings of the NEF and biannual meetings of the British Horse Council do help maintain some sense of a sector. Robert Huey, the Chief Veterinary Officer in Northern Ireland, commented at the 2022 National Equine Forum that the industry in Northern Ireland remains very fragmented. He bemoaned the fact that there are 35,000 horses and over 30 organisations representing them.[8]

In a 1980 analysis of the politics of horse-racing, political scientist Christopher Hill argued that the relationships between the many organisations involved in the politics of racing were becoming "increasingly harmonious".[9] This was in comparison to the preceding period when there were tensions between the Jockey Club and Levy Board in particular over the arrangements and operation of the Levy. Over the twentieth century, governmental interest in the horse shifted from a concern about sufficient horsepower for military and, to a lesser extent, agricultural needs, to one where the horse industry is valued principally for the socio-economic benefits it brings, but with little need seen for governmental interference. As the Government's interest in the horse industry has dwindled, so did the need for a clear and coherent voice for the industry to engage with government lessens, and a more fragmented network of representative bodies is tolerated, each pursing its own specialist interests and seeing less need for a clear and coherent voice representing the sector as a whole. When national crises strike, such as the looming threat of a no-deal Brexit or the Covid pandemic, the value of a coherent and unified horse sector engaging with government departments and agencies becomes clearer.

The Social Complexion of Equestrianism

The horse sector sometimes suffers from the perception that it is the exclusive preserve of the affluent, an elite pastime. Certainly, horse-racing is known as 'the sport of kings', and other equestrian sports also enjoy close associations with the heights of royal and aristocratic society. Queen Elizabeth II was known for her love of horses. She was a racehorse owner and breeder and took a keen interest in her horses at Sandringham and Windsor.[10] She was reported to have an encyclopaedic knowledge of horse-racing. In 2021, the then Duchess of Cornwall said of the Queen's interest in horses,

> I think this is her passion in life, and she loves it and you can tell how much she loves it. … She can tell you every horse she's bred and owned, from the very beginning, she doesn't forget anything. I can hardly remember what I bred a year ago, so she's encyclopaedic about her knowledge.[11]

King Charles was an enthusiastic foxhunter and polo player as a young man. The Princess Royal, Queen Elizabeth's daughter, competed at the Olympics in 1976 with the British equestrian team, and has long maintained an interest in the horse industry. She is currently the President of the NEF. Among the younger generation royals, the Prince of Wales and his brother, the Duke of Sussex, have both been associated with polo. Horses feature heavily in popular narratives and understandings of the royal family, and members of the royal family are prominent features of popular understandings of the horse world. The horse world has long been associated with British society's upper echelons and is also a world in which the socially aspiring can affirm and reinforce their social identity and social standing.

Owning, breeding and training elite horses for racing and other equestrian sports is an expensive business. Even owning a single horse is costly. Just to maintain a horse at home (including feed and necessary maintenance) would typically cost several hundred pounds per month if all the work was done by the owner. If a horse is kept in a livery, then the typical cost will be more. This restricts horse ownership to those households with sufficient means to afford such costs. However, not all riders own horses, and there are more than twice as many regular riders (who ride at least once a month) as there are horses. The surveys conducted by the British Equestrian Trade Association track participation in horse-related activities but also details of the social complexion of participants. Interest in horse-related activities is balanced between the sexes with an almost 50–50 split. However, there are greater differences within particular activities. For example, those with an interest in racing are more likely to be male, but those who ride horses are much more likely to be female. The majority of regular riders are female, although the proportion of regular riders who are male has increased from 23 per cent in 1999 to 38 per cent in 2023.[12] Interest is also spread across age groups, with 15 per cent of regular horse riders aged over 45 and 22 per cent under 16. For those with an interest in horses, the social profile is broadly the same as that for the whole population. Overall, 26 per cent of those with an interest in horses were in the AB socioeconomic group, 28 per cent were in C1 and 46 per cent were in C2DE.[13] Horse ownership is more skewed to the better off. A survey in 2015 found that 30 per cent of horse owners were in the AB socioeconomic group, 34 per cent were in C1 and 36 per cent were in C2D2.[14]

Anthropologist Rosalie Jones McVey has studied the role of equestrianism in the social identities of horse riders. She found the typical horse enthusiast to be a middle-class woman for whom 'horsiness' is felt to be part of her true nature.[15] In contrast to Rebecca Cassidy's earlier ethnography of Newmarket, which found equestrianism was based in a stable class and gender hierarchy rooted in breeding and heritage, Jones McVey found non-racing horse riders developed 'horsey' identities which gave opportunities to "both reflect and reinscribe the essentialism of broader societal gender norms and expectations".[16] Although seemingly symptomatic of post-industrial middle-class conspicuous consumption, her subjects saw 'horsiness' "not as a lifestyle choice … but as part of an individual's true nature or even soul",[17] as something that differentiates them from the 'non-horsey' world and rooted in a "capacity for a particular form of relatedness (with horses)".[18] Jones McVey's horsey women considered themselves "tougher, more resilient, independent and less image obsessed than non-horsey women" and crucially celebrated a commitment and lifestyle that had been actively chosen.

Much is made by those in the horse sector about the cross-class complexion of participants and those involved in the worlds of horses. Race meetings are occasions where people from all sorts of social backgrounds may come together united by a common interest in the horse and the sport.[19] Certainly, people on more modest incomes can be found working in the horse industry. Grooms and stable hands, for instance, are intimately rooted in the world of horses but are usually paid modest incomes. Ahead of the rise in the National Minimum Wage from £8.91 to £9.50 per hour for those aged 23 and above, in April 2022, the Equestrian Employers' Association conducted a survey of its members. Among responses from around 100 businesses, 41 per cent said they would have to make changes to, or lose staff from, their workforce in order to accommodate the increase and 71 per cent were clear that their business would not cope with any further increase.[20] The Low Pay Commission has identified the equestrian industry as one where there are particular problems of low pay.[21] The 2022 National Equine Forum heard a report of an employee who had been working in an eventing yard since January 2018. She had been told that she was self-employed on starting the job, had no contract and was being paid in cash. She was working 6 days a week, officially from 7:30 am to 5 pm, but she had so much work to do that she never finished on time. Her accommodation was charged at £130 per week and she had no paid annual leave. With overtime considered, the British Grooms' Association calculated that by March 2022 she was effectively owed around £12,000 in unpaid wages. The survey found that almost a half (47 per cent) of grooms had not been given a written contract.[22] Poor pay and employment conditions have become seen as an increasing problem in the horse sector. Attracting and retaining staff to work in the horse industry has become a prominent challenge for businesses such as riding stables, liveries and other businesses looking after horses because of the low pay and because of labour market pressures since Brexit.

The social complexion of the horse world might be more diverse than it may be perceived by those outside. It can be characterised by critics as elitist, and at times the horse world gets drawn into wider public controversies which in turn attract attention to its own social make-up. One high-profile saga was the hunting debate around the efforts to ban hunting with hounds during the late 1990s and early 2000s. Through the 1970s and 1980s, the public debate about the rights and wrongs of hunting wild mammals such as foxes with hounds had centred on questions of animal welfare and pest control. In other words, was hunting unnecessarily cruel to the quarry species, and was it necessary at all as a form of controlling pests? In the 1990s, defenders of hunting with hounds also argued that hunting was part of a valued rural tradition that was misunderstood by a naïve and remote urban population, that there was significant economic value in the broader 'hunting economy' and that banning hunting would cost jobs and damage the rural economy. The then Chief Executive of the Countryside Alliance, Janet George, was subsequently quite explicit about the campaigning strategy, "Wrap up hunting in the wider rural fabric. Because everyone loves the country and hates hunting".[23]

Equestrian organisations were involved in the campaign to save hunting with hounds. The Hunting Inquiry, chaired by Lord Burns, received written evidence from more than 30 equestrian organisations such as the Association of British Riding Schools, the British Equestrian Trade Association, the BHIC and the British Horse Trials Association.[24] Albion Saddlemakers based in Walsall estimated that a hunting ban would result in the loss of between 50 and 60 per cent of their workforce of 50 people. The Association of British Riding Schools argued that a hunting ban would "add yet another nail in the coffin" and most certainly lead to further unemployment. The Director of the Badminton Horse Trials reported that a hunting ban would have "a catastrophic effect" on his sport. The Managing Director of the British Bloodstock Agency submitted that "there would be a decline in the breeding of good National Hunt horses and a consequent decline in yet another rural industry and its associated employment". The British Equestrian Trade Association forecast quite specifically that £47.6 million and 45,000 horses "would go out of the economy over 4 years" and that "the total loss to the industry will be £179.4 million and 7,775 jobs".[25] The equine organisations almost in their entirety expressed concerns or opposition to a ban on hunting with dogs and emphasised the economic losses to the industry. Central to these claims were two critical assumptions. First was that if hunting live quarry with hounds were banned, hunts would not continue to ride to hounds. They would make no attempt to find new ways to continue to follow a scent, enjoy the chase but avoid the kill. The second assumption was that horses used for hunting would no longer be used at all and so all the expenditure on their upkeep would be lost to the economy.

A reading of these submissions over two decades later is striking not only for the catastrophic scenarios rehearsed by the equestrian organisations, but also for the commonality of arguments and almost unified voice. The submissions give a strong sense of an entire industry opposed to a ban. Opinion polling by Mori conducted in 1997 found that 78 per cent of the population opposed hunting with dogs while 8 per cent were in favour. Among those living in rural areas, 63 per cent were opposed while 21 per cent were in favour. Notably, among horse riders, 51 per cent were opposed while 35 per cent were in favour. Although opposition to hunting was weaker among horse riders than the general population, it was still the case that a greater proportion of horse-riders polled opposed hunting with dogs than those who supported it.[26] Further polling in 1998 after several months of intense public and political debate about the issue found that 68 per cent of horse riders now supported the ban.[27] This is in contrast to the submissions by the industry representative bodies to the Inquiry. After the legislation was passed and hunting with dogs was eventually banned, hunts adapted their practices and were able to continue to organise hunts and ride to hounds. It was widely reported in the years after the ban came into effect that levels of participation in hunting increased, with rural businesses dependent on hunting activities reporting a marked increase in trade.[28]

There have been efforts to widen participation in horse activities and break away from the perception of elitism. However, controversies have continued to dog parts of the horse sector. For example, in July 2017 Clare Salmon, then Chief Executive of the British Equestrian Federation (now British Equestrian), resigned from her position complaining that she was the victim of bullying and that the organisation suffered from a toxic culture of elitism. An independent investigation was conducted into her complaint and the outcome was published in March 2018. Referring to her accusation of elitism, she was quoted as saying:

> At one of the board meetings that I attended with one of my colleagues, we said that we wanted to open up horse sport to a wider audience and we were told: "But just anybody might turn up, think who might turn up, how terrible might that be" – people who were beyond the sort of tweedocracy of traditional horse sport.

She had set out a vision for widening participation in the sport but became frustrated that she was being hampered and undermined when implementing this vision by more conservative forces within the organisation. She publicly expressed her distress at the episode and said she had been trolled online. The independent investigation found that she had been the victim of bullying, concluding that "certain actions can, in the panel's opinion, be objectively viewed as bullying, elitist and arising from self-interest but not corruption". Commenting on the findings, Clare Salmon said:

> I think I am both relieved and sad in reaction to the report. I am very relieved that the report has vindicated the concerns I raised about bullying, about elitism and about a toxic culture within the equestrian world. I am also very sad because I am a passionate advocate of horses as a force for good and I still believe in the vision that I articulated at the time and that the British Equestrian Federation are now attempting to implement.[29]

There has been some debate and self-consciousness about the social complexion of the horse world, and it is undoubtedly the case that those who ride and use riding schools are from a more diverse spread of the income range than those who own horses. British Equestrian commissioned research in 2022 into the barriers inhibiting participation in equestrian sports.[30] There has been less open discussion and self-reflection on the ethnic complexion of the horse world. It is extremely white, with a marked under-representation of people from ethnic minority backgrounds. Some 18 per cent of the UK population are from Black, Asian or mixed ethnic backgrounds. British horse society is reflected in the pages of *Horse & Hound*, the oldest and largest selling horse magazine with a weekly circulation of around 45,000. The magazine has contained some coverage of the need to improve equality, diversity and inclusion in the horse world. However, a survey of a random sample of six editions of the magazine in 2022 revealed only three non-white people among over 770 various horse riders, award winners and journalists featured in its pages (less than 0.4 per cent), or a proportion 47 times smaller than for the British population as a whole.[31]

Horses, Welfare and Wellbeing

Philosopher and social reformer, Jeremy Bentham, used the sentience, intelligence and personability of the horse in his foundational argument about animal welfare and animal rights. In 1798 he wrote:

> a full grown horse or dog is beyond comparison a more rational, as well as a more conversable animal, than an infant of a day, or a week, or even a month old. But suppose the case were otherwise, what would it avail? the question is not, Can they *reason?* nor, Can they *talk?* but, Can they *suffer?*[32]

Since Bentham"s time, animal welfare issues have come to have a greater influence over how the horse is thought about and treated. The rise in profile of animal welfare campaigning since the 1960s has been a feature of post-industrial societies generally but has been particularly notable in the UK.[33] Membership of the Royal Society for the Protection of Cruelty to Animals grew from

4,000 in the 1930s to 770,000 in 1990.[34] It now has more than a million regular financial contributors. Concern about the welfare of the horse is not a wholly new phenomenon. Indeed, in Victorian Britain there were concerns about the treatment of working horses in towns in particular.[35] However, the growing influence of animal welfare values has meant that horse welfare has become an increasingly prominent feature of the work of equestrian organisations over recent decades.[36]

The animal welfare implications of horseracing have been a cause of some controversy. The League Against Cruel Sports believes that races of four miles or more involving several fences are too gruelling for horses, and they draw attention to fatalities at horseracing events, especially high-profile events such as the Grand National and Cheltenham Festival. They protest that "hundreds of horses die on the racecourse every year".[37] Animal Aid has monitored racing fatalities through its 'Racehorse Death Watch' website.[38] The website contains an annual report on the number of injuries and deaths with figures compiled since 2007, with over 2,600 horse deaths since that year. In the first year of gathering data, Animal Aid recorded 161 deaths of racehorses on racecourses, 137 of which were at National Hunt events (over jumps).[39] Typically, the organisation reports between 130 and 220 deaths of racehorses per year in the UK because of accidents during horseracing events, with an average of around 160. In 2019, of 186 deaths, 145 were in National Hunt racing and horse deaths occurred at 52 different racecourses. There is little sign of any improving trend over the period 2007 to 2019, suggesting this level of horse deaths is to be generally expected in the sport. In April 2023, the start of the Grand National was delayed as a result of protests by animal welfare campaigners, and over 100 arrests were made. Three horses died during the festival at Aintree.

Animal Aid also campaigns against the general treatment of horses during racing, including the practice of whipping which they argue contributes to horse deaths and injuries because it cajoles horses to run even faster and at greater risk. Whipping has been the subject of its own public controversy for some time and the sport faces pressure to tighten its rules.[40] Animal Aid argues that the financial imperative to win increases the risk to horses, and the economic incentives around the sport mean racing takes place in conditions that may be too dangerous. It argues that the body responsible for welfare in horseracing, the BHA, is compromised by its interest in the economic development of the sport.[41] There were six horse deaths at the 2018 Cheltenham Festival which caused a public furore and forced the BHA to conduct a review of the event from a horse welfare perspective.[42] A UK Parliament Petition organised by animal welfare campaigners attracted over 100,000 signatures and called for an independent welfare regulator, an idea which was included in the 2019 General Election manifesto of the Liberal Democrat party.

In 2019, the BHA established a new Horse Welfare Board and in 2020 published a new strategic plan for the welfare of horses in racing.[43] The Board was established as an independent body and contains members from the BHA and the racing industry but no members from animal welfare organisations. (Animal

Aid said "With industry representatives dominating the Board, it is merely a halfway house to independent regulation".[44]) The Board argued that animal welfare is "an emotive issue", supported the principle of self-regulation, but called for a more collaborative approach. In assessing the social context for current animal welfare concerns about the horse, it argued that young people are becoming more remote from the countryside and horses are rarely seen in cities.

> This modern lack of familiarity with, and understanding of, the horse, gives rise to myths and misperceptions. People may assume, for example, that a horse turned out in a field will be happier and safer than a working horse, whether that working horse is a police or military horse, or a sport horse, whereas the opposite may well be true. Racing must work together, along with other equine sports and sectors, to grow public understanding of, and exposure to, horses, to manage this gap in perception and understanding, particularly amongst young people and in urban communities.[45]

The response of the sport's governing bodies to animal welfare concerns has echoes of the charge levelled by Jonathan Meades in the 1990s that racing is administered "with a patrician hauteur".[46] There has also been a growing campaign against perceived poor treatment of horses in international equestrian sport, and calls for equestrian sport to be dropped from the Olympic Games.[47] While injuries to racehorses can be catastrophic and very public, the damage done through Olympic equestrian disciplines can be more insidious. A study in the *Equine Veterinary Journal* found that 9.3 per cent of horses examined at equestrian competitions had oral lesions or blood in the corners of their mouths after competing, for example.[48]

Beyond the higher-profile welfare controversies around horseracing, animal welfare has been an issue of more general concern across the horse sector. The Royal Society for the Prevention of Cruelty to Animals (RSPCA) estimate that there are around 7,000 horses in England and Wales in welfare need and most equestrian organisations have given heightened attention to animal welfare issues over recent years. An independent review published in 2016 found almost 40 different areas of animal welfare concerns about horses and noted that many of the problems highlighted had not been the focus of much scientific investigation.[49] More recently, evidence has emerged about the problem of obesity and horses being overweight. Among leisure horses in the UK, it is estimated that between a third and a half (31 per cent to 54 per cent) of the population may be affected.[50] This reflects a problem with companion animals more generally, and there is evidence of increasing obesity among cats and dogs too. The study of companion horses found that "owners find it hard to tell when their horse is overweight; and that even when they do realise the horse is overweight, it is difficult for them to alter their horse's weight within the UK's obesogenic environment".[51]

Horses can play a role in schemes to improve human wellbeing. For example, care for, and work with, horses commonly featured in the therapeutic and rehabilitative activities of prisoners who worked on Britain's prison farms well into the twentieth century.[52] More recently, equine-assisted therapy and learning is increasingly used as a therapeutic benefit for those people who find it difficult to engage with conventional therapeutic interventions or practices. These may be psychological therapies or forms of experiential learning to develop specific skills in social, emotional or behavioural development. Two organisations, the Professional Association of Therapeutic Horsemanship International and the Equine Assisted Growth and Learning Association, developed in the United States and an international body, Horses in Education and Therapy, operates across several countries. The Equine Assisted and Facilitated Practitioners Network (EAFPN) was established in the UK in 2013 and had attracted over a thousand members within 5 years.[53] (Its Facebook page currently has over 3,500 members.) Animals have been used in therapy for centuries but had not been documented as a formal practice until the 1960s with the use of therapy dogs.[54] Horses are thought to bring particular benefits because of the vulnerabilities in their nature as prey animals, their sociability and highly developed non-verbal communication skills. Their sensitivity and responsiveness to the behaviour of others, including humans, mean they can provide feedback which supports therapeutic processes. The skills that are needed to build relationships with horses are transferable to the skills people require in daily life. These include being assertive, setting boundaries, good communication, creativity and resilience. For people suffering from depression or problems of low self-esteem, it can be a positive and therapeutic experience to engage with and build a relationship with a horse.[55]

In a case study of working with young people aged 11 to 15 years who had suffered trauma, abuse or neglect, Hannah Burgon and colleagues researched how participants were able to step into a new relationship with a horse which then helped facilitate a positive relationship with a therapeutic practitioner.

> Through modelling real, congruent, respectful and caring relationships with the horses which is demonstrated by acknowledging their different needs and personalities and can also include setting boundaries and sometimes confronting difficult decisions, participants feel safe that their needs will also be respected.[56]

The approach met with overwhelming positive feedback from participants and carers. Equine-assisted therapy can be used to address issues of addiction, anger, anxiety, depression and trauma.

Horses and Social Control

Horses continue to be used in maintaining public order and social control. This is typically through the police horse which has attracted very little attention in the academic literature on horses. While mounted forces have been used to maintain peace and public order for centuries, it was the establishment of

London's Bow Street Police Horse Patrol, which began its work in 1763 with eight horses, that represents the first mounted police force. (Mounted police units were subsequently established in Australia and Texas in the 1820s.) By the early nineteenth century the unit had become "an integral cog of the Metropolitan Police".[57] Horses became increasingly used for crowd control during the nineteenth century and were frequently involved in responding to the militant suffragette movement in the early twentieth century. By the 1960s, the metropolitan mounted branch included 210 officers and 201 horses. Units were similarly built up in other major British cities and despite the use of police cars and vans, mounted units and teams of police horses were maintained. In the 1980s, there were 18 UK police forces with mounted branches, with a total of 421 horses. Of these, 176 were in London, 27 were in Liverpool, six in Newcastle, and six in Edinburgh.[58] By 2014, the number of units in the UK had fallen by a third to just 12, involving a total of 271 officers and 247 horses. The West Midlands Police disbanded their mounted unit in 1999. Essex Police and Nottinghamshire disbanded theirs in 2012. Cleveland and Humberside Police both followed in 2013.[59] In 2016, the Metropolitan Police continued to operate 110 police horses across London,[60] but by 2023, the number had fallen to just 38.[61]

Very little has been written about mounted police and the continuing role of horses in managing urban public order. In a famous incident in 1923, a crowd covering the pitch at a Wembley FA Cup final was cleared by a white horse named Billie, ridden by a police constable. In more recent times, police horses have dominated television images of the policing of the miners' strike in the 1980s and the poll tax riot of 1990. Mounted police can represent both a reassurance and a threat to the public. They are intended as a calming presence signifying order and control, but at the same time they embody a threat of physical force. They are primarily used for routine patrol work and have enjoyed high levels of positive support as an approach to urban policing, especially in London, even in the age of the motor vehicle. In controlling crowds, police can also use dogs, but horses are seen as having a stronger, calming and reassuring effect because of their docile nature. However, a couple of high-profile public order controversies in the 1980s and 1990s re-cast popular images of the police horse from the calming and ceremonial roles on city streets to a much more violent and militaristic experience more evocative of Peterloo.[62]

Acute challenges in managing public order arose during the miners' strike of the 1980s when secondary picketing was prohibited and police became embroiled in violent confrontations with striking miners and their supporters. On 18 June 1984, a demonstration took place at a coking plant at Orgreave in South Yorkshire where a crowd of several thousand pickets were confronted by around 3,000 police in one of the largest peacetime policing operations in the UK.[63] Miners' leader Arthur Scargill had called for mass picketing of the plant to disrupt the production of coke for use in steel production. A police manual produced by the Association of Chief Police Officers and subsequently disclosed at a trial of defendants accused of rioting at Orgreave provided guidance on tactical operations. In deploying horses it proposed manoeuvres to intimidate protestors, talking of "dispersing a crowd using impetus to create fear" and how mounted officers should "advance on a crowd in a way indicating that they do not intend to stop", ideally at a canter. It explained how mounted and foot police could work together:

> ... the mounted police officers form in a double rank, line abreast facing the crowd and advance together at a smart pace towards the crowd. Foot officers stand well aside to let them through and reform behind following at the double.[64]

At Orgreave, police behaviour was subsequently widely criticised. Several mounted charges were deployed against the pickets as lorries were arriving at the plant. After the protest had quietened down and many pickets had left, the mounted police charged again, and pickets were chased through the village streets by mounted police. Ninety-three arrests were made, and 51 pickets and 71 police were injured in the battle. Charges of rioting were pursued to court, but the trials collapsed, and South Yorkshire police later agreed to pay a total of £425,000 and £100,000 in legal costs to 39 pickets in an out-of-court settlement. The episode has been described as "the most violent police behaviour ever seen in a modern industrial dispute",[65] and a campaign continues for an official inquiry into the events at Orgreave. The episode has gone down in British industrial folklore. It was attended by an Economic and Social Research Council-funded postdoctoral researcher studying public disorder who more than 30 years later wrote recalling "the sinister reverberation of the horses' hooves as they close[d] in on the pickets".[66] The episode has been called a 'police riot', and the researcher, now Professor David Waddington, later wrote:

> I saw a mounted police officer chase a group of pickets right up to the entrance of the nearby Asda supermarket. As the latter were about to disappear through the automatic sliding doors, the mounted policeman ducked down with the apparent intention of pursuing them into the busy shopping area. Following two more abortive attempts, he clearly thought better of it, turned and cantered off in search of easier forms of quarry.[67]

Media coverage of the event presented a narrative of the police responding to being attacked by pickets, but the campaigners, including prominent lawyers such as Gareth Pierce and Michael Mansfield, have continued to assert that the police strategy on the day appeared to be to attack the pickets on horseback.

A later high-profile episode of public disorder involving police horses took place on 31 March 1990. A march of an estimated 200,000 people processed from Kennington Park to Trafalgar Square in protest at Mrs Thatcher's proposals to reform local government finance by introducing a poll tax. The tax had provoked widespread public opposition, and many of the marchers were taking part in a public demonstration for the first time. The procession was largely peaceful, but disturbances broke out in Whitehall and Trafalgar Square towards the end of the march. Some accounts have the use of mounted police charges as the trigger for the rioting. A man in a wheelchair was arrested and treated roughly by police, and a woman was arrested and strip-searched in front of the crowd. Demonstrators began to sit down in Whitehall. "Mounted riot police baton-charged the crowd. The crowd, angered by this violent provocation, retaliated throwing sticks, banner poles, bottles – anything they could find".[68] Tensions continued, and buildings were set alight around Trafalgar Square. Police vans were driven at speed through the crowd which further enraged some demonstrators. The rioting, which had been "unpredicted and unexpected", later spread through London's West End.[69] Police horses were used in repeated mounted charges into the crowd to disperse protestors.[70] National television news included shocking violent footage of protestors being trampled under the police horses' hooves. The riots represented a failure of policing in managing public order in a large demonstration, but more than this it was the police's misreading of crowd psychology and heavy-handed crowd-control measures, including charging by police horses, that helped ignite and inflame the disorder.[71]

After 2010 and the onset of austerity, questions began to be asked about the role and value of mounted horses in policing. A national study was carried out to examine what mounted police were used for, commissioned by a working group of senior police officers from forces that used horses.[72] Of the almost 40,000 deployments in 2013, the largest number was for local high visibility patrols in neighbourhoods (60 per cent). Public order events including football matches accounted for almost 5,000 deployments (or 15 per cent), while ceremonial events accounted for 9 per cent. It found that mounted police were less involved in direct crime detection and reduction than other types of police officers.

The use of mounted police is informed by socio-psychological literature on crowd theory and the 'Elaborated Social Identity Model' in particular. The model addresses the underlying socio-psychological processes that influence crowd dynamics and responses to police activity.[73] It stresses the principles of understanding the motives and sensitivities in play within a crowd, strategically facilitating legitimate intentions, increasing capacity for communication between police and crowd members and differentiation of crowd members to avoid indiscriminate coercion. Researchers carried out detailed observational work to assess the effectiveness of mounted police in different types of policing contexts. They found generally positive receptions in neighbourhood policing, including at large events such as Glastonbury Festival. Positive effects at football matches were less discernible and clear cut. Focus group work and other research found that "the image of mounted police intervening in public protest can have lasting consequences for the public memory of policing".[74] The study concluded:

> the effect and effectiveness of mounted police in crowd settings will be intertwined with how their use is understood by, and communicated to, crowd members. While it is not always possible to provide advance communication

> on what police will do in a crowd setting, where such communication is possible it would appear valuable to include messaging about mounted police, and this may impact on the overall effectiveness of their deployment.[75]

Overall, the study concluded that mounted police are a unique policing resource with both heightened response and public engagement value. Based on its analysis of use patterns and demonstrable value of mounted police units, it argued that consideration should be given to positioning them strategically as a resource primarily to support neighbourhood policing.

Conclusions

The horse world may seem at first glance to be an exclusive preserve of the better off. Charges of elitism and exclusivity are easy to make. The Royal Family and the ruler of Dubai may love their horses, and the British Equestrian Federation may be concerned that efforts to broaden participation might let in riff-raff. However, people who ride and enjoy horses are more socially diverse than initial stereotypes might imply, at least when measured by social class. Many more people regularly ride horses than own horses, and there are almost five times as many regular riders as there are horse-owning households. Furthermore, there are three times as many people who watch horseracing as there are horse riders. The horse world in its broadest form embraces a wide range of people, and in Britain the population still touched by the horse in some way spans tens of millions. Nevertheless, the horse world remains highly disproportionately white, and certainly so in the way it represents itself to itself through its most popular publication, *Horse & Hound*.

The institutional governance of the horse world is complex, even after the efforts to bring the world together to speak with one voice of two decades ago. One key cleavage is between the elite horseracing world and the rest. Racehorse ownership requires a certain level of wealth. The sport is highly internationalised and appeals to a global elite, as well as the regular punter, and enjoys a continuous injection of public funds to sustain the industry by means of the Horserace Betting Levy. Beyond racing, the horse world is represented by an array of sporting, business and animal welfare organisations, bodies representing riding schools, rare breed enthusiasts and a growing number of charities interested in how horses enhance human wellbeing. If we consider the evolution of the preoccupations of the horse world over the past two or three decades, then the dominant trend is the growing influence of animal welfare concerns. Many equestrian organisations lined up behind the argument that banning hunting with hounds would cause untold damage to the rural economy. In the two decades since the ban, animal welfare organisations and campaigners have engaged more extensively with equestrian sports. As equestrian veterinary science evolves and our understanding of horse health, wellbeing and behaviour develops, animal welfare comes to impinge more and more on the horse world, and established practices in equestrian sport and general care for horses get called into question. It is to the evolving knowledge and understanding of horses and their interactions with humans, and its implications for the horse world, that we turn in Chapter 8.

Notes

1 British Horseracing Authority (2020b) *Annual Report and Consolidated Financial Statements*. London: British Horseracing Authority.
2 Woodhouse, J. (2019) *The Horserace Betting Levy*. House of Commons Library Briefing Paper No. 7368. London: House of Commons. See also Horserace Betting Levy Board (2023) *Annual Report and Accounts 2021/22*. London: Horserace Betting Levy Board.
3 Crossman, G. (2010) *The Organisational Landscape of the English Horse Industry: A Contrast with Sweden and the Netherlands*. University of Exeter PhD thesis, p.142.
4 PWC (2018) *The Contribution of Thoroughbred Breeding to the UK Economy and Factors Impacting the Industry's Supply Chain*. Newmarket: Thoroughbred Breeders Association.
5 Crossman (2010).
6 Donoughue's two volumes of political diaries of his time as a minister detail his battles and frustrations with the senior civil servants in the Ministry and especially Permanent Secretary Richard Packer. See Donoughue, B. (2016) *Westminster Diary Volume 1: A Reluctant Minister Under Tony Blair*. London: I.B. Tauris; Donoughue, B. (2018) *Westminster Diary Volume 2: Farewell to Office*. London: I.B. Tauris.
7 Crossman (2010), pp.128–33.
8 Huey, R. (2022) What's the value of a horse?: An economic strategy for the equine sector in Northern Ireland, presentation to the National Equine Forum, 3rd March 2022.
9 Hill, C. (1988) *Horse Power: The Politics of the Turf*. Manchester: Manchester University Press, p.249.
10 Pitt, M. (ed.) (2016) *The Queen and Her Horses*. London: Horse & Hound.
11 Murphy, V. (2021) Queen Elizabeth Has Encyclopedic Knowledge of Horse Racing And it's Her Passion in Life, Camilla Says, *Town & Country Magazine*, 15 June. See also: Pitt, M. (ed.) (2016) *The Queen and her Horses*. London: Horse & Hound.
12 JDA Research (2023) *The National Equestrian Survey 2023. Overview Report*. Wakefield: British Equestrian Trade Association, p.12.
13 The Henley Centre (2004) *A Report of Research on the Horse Industry in Great Britain*. Report to Defra and the British Horse Industry Confederation. London: Defra, pp.42–44.
14 Two Circles (2015) *The National Equestrian Survey. 2015. Overview Report*. Wakefield: British Equestrian Trade Association, p.22. Socio-economic categories A and B are higher and intermediate managerial, administrative or professional occupations. C1 are those in supervisory or clerical and junior managerial, administrative or professional roles. C2 are skilled manual workers. D are semi- and unskilled workers.
15 Jones McVey, R. (2018) *Reasonable Creatures: British Equestrianism and Epistemological Responsibility in Late Modernity*. University of Cambridge PhD Thesis, p.34.
16 Jones McVey, p.75.
17 Jones McVey, p.42.
18 Jones McVey, p.47.
19 Fox, K. (1999) *The Racing Tribe: Watching the Horsewatchers*. London: Metro Books.
20 Equestrian Employers' Association (2022) 71% concerned about viability, *News Release* 3rd March 2022 https://equestrianemployers.org.uk/news/546/71-concerned-about-viability

21 Low Pay Commission (2020) *National Minimum Wage*. London: Low Pay Commission, p.199.
22 Katan, L. (2022) A view from the ground up, presentation to the National Equine Forum 3 March 2022. https://www.nationalequineforum.com/forum-2022/
23 Quoted in Ward, N. (1999) Foxing the nation: the economic (in)significance of hunting with hounds in Britain, *Journal of Rural Studies* 15, p.401; see also J. George (1999) *A Rural Uprising: The Battle to Save Hunting with Hounds* London: J. A. Allen.
24 Committee of Inquiry into Hunting with Dogs in England and Wales (2000) *Report of the Committee of Inquiry into Hunting with Dogs in England and Wales* Cm.4763. London: Home Office.
25 The quotations in this paragraph are all drawn from the written submissions of evidence to the Committee of Inquiry into Hunting with Dogs in England and Wales (2000).
26 Leaman, J. (1998) *Rural Opinion: Flushing Out the Truth*. Presentation to the Countryside Protection Group Conference, London, 24 February, London: MORI.
27 Campaign for the Protection of Hunted Animals (1998) Anglers and riders support Foster Bill to ban hunting with dogs, *Campaign for the Protection of Hunted Animals Press Release* 3 February, London: CPHA.
28 See, for example, Taylor, M. (2007) Hunt supporters claim legislation has backfired, *The Guardian* 27 December 2007, p.7; Townsend, M. (2008) Hunting ban sparks a rural boom, *The Observer* 24 February 2008, p.26; Harvey, F. (2013) Huge rise in first-time hunters for new foxhunting season, *The Guardian* 2 November 2013, p.13. I experienced some harassment for publishing research that called into question the hunt supporters' claim that hunting legislation would be catastrophic for the rural economy. However, the argument I then made can now be tested against the experience of the two decades since the 2003 legislation – See Ward, N. (1999) Foxing the nation: the economic (in)significance of hunting with hounds in Britain, *Journal of Rural Studies* 15, 389–403.
29 BBC Sport (2018) British Equestrian: Chief Executive Clare Salmon 'suffered bullying and elitism in role' 15 March 2018, https://www.bbc.co.uk/sport/equestrian/43409111
30 British Equestrian (2022) *Barriers to Equestrianism: Invitation to Tender*. Kenilworth: British Equestrian.
31 The six randomly sampled editions were 16th June, 28th July, 1st September, 20th October, 17th November and 22nd December. Two of these editions (16th June and 28th July) contained articles on the issue of equality and diversity and under-representation of non-white people in the horse sector.
32 http://www.animal-rights-library.com/texts-c/bentham01.htm
33 Garner, R. (1993) *Animals, Politics and Morality*. Manchester: Manchester University Press; Garner, R. (2002) Political science and animal studies, *Society and Animals* 10, 395–401.
34 Garner (1993), p.44.
35 James, J. (1994) *Debt of Honour: History of the International League for the Protection of Horses* Basingstoke: Macmillan.
36 James (1994).
37 League Against Cruel Sports (2021) Horse racing, https://www.league.org.uk/what-we-do/protect-animals/horse-racing/
38 Race Horse Death Watch https://www.horsedeathwatch.com/reports.php
39 Animal Aid (2008) *Race Horse Death Watch: The First Year*. Tonbridge: Animal Aid.
40 Jones, B. *et al.* (2015) A critical analysis of the British Horseracing Authority's review of the use of the whip in horseracing. *Animals* 5, 138–150.
41 Animal Aid (2020) *186 Race Horses killed in 2019*. Tonbridge: Animal Aid.

42 British Horseracing Authority (2018) *The Cheltenham Festival Review 2018*. London: British Horseracing Authority.
43 Horse Welfare Board (2020) *A Life Well Lived: A New Strategic Plan for the Welfare of Horses Bred for Racing 2020–2024*. London: British Horseracing Authority.
44 Animal Aid (2020).
45 Horse Welfare Board (2020), p.27.
46 Meads, J. (1997) *Even Further Abroad – Nag, Nag, Nag*, BBC Two.
47 Taylor, J. (2022) *'I Can't Watch Anymore': The Case for Dropping Equestrian from the Olympic Games*. Copenhagen: Epona Publishing.
48 Uldahl, M. and Clayton, H. (2019) Lesions associated with the use of bits, nosebands, spurs and whips in Danish competition horses, *Equine Veterinary Journal* 51 154–162, quoted in Taylor (2022), p.132.
49 Horseman, S. *et al.* (2016) Current welfare problems facing horses in Great Britain as identified by equine stakeholders, *PLoS One* 11, e0160269.
50 Furtado, T. (2021) Exploring horse owners' understanding of obese body condition and weight management in UK leisure horses, *Equine Veterinary Journal* 53, 752–62.
51 Furtado *et al.* (2021), p.759; see also Furtado, T. *et al.* (2021) Hidden in plain sight: Uncovering the obesogenic environment surrounding the UK's leisure horses, *Antrozoös* 34, 491–506.
52 Wright, H. (2017) *Outside Time: A Personal History of Prison Farming and Gardening*. Eightslate: Placewise Press.
53 Burgon, H. *et al.* (2018) Hoofbeats and heartbeats: equine-assisted therapy and learning with young people with psychological issues – theory and practice, *Journal of Social Work Practice* 32, p.3; see also Burgon, H. (2011) *A Reflexive Exploration of the Experiences of At-Risk Young People Participating in Therapeutic Horsemanship*. PhD Thesis, Cardiff University School of Social Sciences.
54 Morrison, M. (2007) Health benefits of animal-assisted interventions, *Complementary Health Practice Review* 12, 51–62.
55 A brief introduction to equine assisted therapy by Danielle Mills of the University of Derby can be found at: https://www.counselling-directory.org.uk/equine-assisted-therapy.html
56 Burgon *et al.* (2018), p.14.
57 Roth, M. (1998) Mounted police forces: a comparative history, *Policing: An International Journal of Police Strategies and Management* 21, p.708.
58 Roth (1998), p.708.
59 Giacomantonio, C. *et al.* (2015) *Making and Breaking Barriers: Assessing the Value of Mounted Police Units in the UK*. Cambridge: RAND Europe, pp.2–3.
60 Question to the Mayor of London, 27th June 2016. https://www.london.gov.uk/questions/2016/2126
61 The figure of 38 police horses in 2023 comes from the response to a Freedom of Information Act request to the Metropolitan Police (No. OC/FOI/23/029204#, 21st March 2023).
62 The Peterloo Massacre took place in Manchester in 1819 when mounted cavalry troops charged into a large crowd of people campaigning for political reform. Fifteen people in the crowd were killed.
63 Roth (1998).
64 Quoted in East, R. *et al.* (1985) The death of mass picketing, *Journal of Law and Society* 12, p.312.
65 Conn, D. (2017) The scandal of Orgreave, *The Guardian* Long Read, 18 May 2017.
66 Waddington, D. (2017) The battle of Orgreave: Afterword, pp.366–74 in R. Page (ed.) *Protest: Stories of Resistance*. Manchester: Comma Press; see also Bedford, M. (2017) The battle of Orgreave, pp.335–65 in R. Page (ed.) *Protest: Stories of Resistance*. Manchester: Comma Press.

67 Waddington (2017), p.371.
68 Burns, D. (1992) *Poll Tax Rebellion*. Stirling: AK Press, p.89.
69 Drury, J. (2017) The poll tax riot – afterword: from peaceful protest to riot, 31 March 1990, pp.422–27 in R. Page (ed.) *Protest: Stories of Resistance*. Manchester: Comma Press, p.423.
70 Waddington, D. (1992) *Contemporary Issues in Public Disorder: A Comparative and Historical Approach*. London: Routledge.
71 Drury (2017).
72 Giacomantonio *et al.* (2015).
73 Stott, C. (2009). *Crowd Psychology and Public Order Policing An Overview of Scientific Theory and Evidence*. Report submitted to the HMIC inquiry into the policing of the London G20 protests. Liverpool: University of Liverpool School of Psychology.
74 Giacomantonio *et al.* (2015), pp.139–40.
75 Giacomantonio *et al.* (2015), p.140.

8 Horse Knowledge

Introduction

Ulrich Raulff writes of the final century of our relationship with the horse. In converting grass and oats into physical force, horses were able to power agriculture, industry and transport through the crucial centuries to industrial take-off. They bore kings, knights and cavalry troops and hauled cannons, carriages and wagons. They helped mobilise whole nations. However, in addition to physical power, horses were also at the centre of systems of knowledge from which humanity benefited. They became a "privileged object of human research and cognition".[1] The horse formed part of several knowledge areas (medical, agrarian, military, artistic) as well as being subject to different approaches to knowledge production (empirical, experimental, literary). Knowledge about horses continues to develop. Advances in archaeology and palaeontology help better understand the evolution and domestication of the horse.[2] Historical research sheds more light on the role of the horse in agriculture, warfare and the development of nations and political systems.[3] Research in the history of art and literature enriches our understanding of the ways we give meaning to the horse.[4]

As the role of the horse shifted during the twentieth century, new scientific questions began to emerge about the horse and our relationships with it. We discussed in Chapter 7 how social pressures around animal welfare have come to have a greater influence over how the horse is thought about and treated. Concerns about horse welfare have prompted new scientific research to reduce harm and promote wellbeing. Horse welfare involves not only the veterinary science of horse physiology and health, but also social and philosophical questions about the ethics of how horses are treated. As concerns about animal welfare in general have intensified over recent decades, the horse has become embroiled in controversies about harms in horseracing and other equestrian sports, about how horses are fed and cared for and about whether it can even be considered ethical to ride a horse.[5]

Horse welfare organisations point to the need to consider the 'social license to operate'. This concept originated in the worlds of mining and oil and gas exploration but has come to be adopted more widely.[6] The term implies the social legitimacy of an enterprise or activity and is usually employed in the face

DOI: 10.4324/9781003454359-8

of some disapproval or controversy. It goes beyond what is strictly legal to mean alignment with prevailing social norms and values, recognising that they may be subject to change over time. Social licence requires layers of active work through first following rules, then establishing approval from stakeholders, then building support through trust-building measures. Social license is not something that can be simply and complacently 'held' but must continually be actively worked on to maintain. The horse world's social licence to operate is under question, and horse welfare organisations urge that the sector thinks in terms of what needs to be done to earn and maintain its social licence in the face of this challenge.

Different realms of scientific (and social science) knowledge will drive change in how we live with and use horses, and new evidence will change the context within which the horse world's social licence is supported and questioned. This chapter considers the different realms of expertise about horses and how they have developed over recent decades. It examines the nature of the equine veterinary profession before exploring recent debates about horse breeding and animal welfare. New scientific knowledge is redefining our understanding of the early domestication of the horse and of equine genetics and breeding. Neuroscience is also transforming understanding of human–horse relationships and the role of the horse in human wellbeing. The chapter explores possible futures for the relationships between humans and horses in the context of developing scientific knowledge about the history of the horse and the directions of contemporary socio-technical change.

Breeding, Sporting Performance and Horse Welfare

Until the 1960s, the term 'animal welfare' featured rarely in public and political discourse. A history of legislating to tackle animal *cruelty* dates back to acts of parliament in the early nineteenth century, but it was only in response to concerns about intensive farming in the 1960s that the concept of animal *welfare* began to gain purchase.[7] Crucially, welfare meant more than just the 'prevention of cruelty' and was increasingly based on a scientific understanding of animal stress, behaviour and wellbeing. In this context, our understanding of horse welfare and wellbeing comes from a mix of sources. Equine vets are in the frontline of horse health, and their professional body, the British Equestrian Veterinary Association, seeks to advance veterinary and allied science, promote scientific excellence in the field and educate equine veterinary professionals around the world.[8] Founded in 1961, the association now has a membership of over 2,800 and publishes two journals, *Equine Veterinary Journal* and *Equine Veterinary Education*. In 2002, following a scientific meeting on horse welfare in Iceland, an International Society for Equitation Science was established. Its first symposium was held in 2005, with international conferences following most years. The Society's mission is "to promote and encourage the application of objective research and consequently advance equestrian practice that will ultimately improve the welfare of horses in their interactions with humans".[9] The Society is a key international forum through which scientific research informs the politics of equine welfare and a vehicle for scientific opinion to take its own stand on welfare issues.

Horse sport has come under increasing scrutiny over recent decades. The dangers and risks are not particularly new, but the spread of information and communication technologies and social media have meant that problems with horse welfare can now be captured and projected to audiences of millions. Racehorses can run at speeds of over 40 miles an hour carrying a rider. Such speeds put the animal at risk of catastrophic injury. When horses suffer serious injuries at the races, they can be put down immediately, but it can be shocking for spectators to witness the rapid transition of an animal from a magnificent athlete to one writhing in agony, and then to a corpse. Those organisations that now collate data on the number of horse deaths at horseracing events are rendering visible and calculable a phenomenon that was previously kept discrete and relatively hidden.[10]

Thoroughbred breeding has its own ethical dilemmas as horses are bred to develop and strengthen their desired attributes of speed and stamina. The genetic trait most responsible for speed can be traced through bloodlines for more than 300 years. Scientific advance has altered the context for horse-breeding, particularly since the publication of the horse genome in 2009.[11] Genetic testing of equine embryos is now available as a commercial service, and there is increasing interest in the question of equine gene editing, although opinion is divided on the acceptability of genetic technologies in equine breeding. On the one hand, testing for inherited diseases could be a useful tool in improving equine health and welfare, and genetic testing for desirable traits may be thought of as an extension of existing selective breeding. However, the effects of gene editing on equine health and welfare are not yet clearly understood. A distinction can be drawn between gene editing for treatment and that for performance enhancement, and there are calls for international co-ordination of regulation and reporting to manage the development of gene editing in horses.[12]

In horseracing there has been considerable focus on the use of the whip to encourage horses to run as fast as possible. Some animal welfare organisations call for the use of the whip to be banned altogether, and a system of rules has developed to regulate the use of the whip and limit the number of times it can be used in a race. The use of the whip has come under pressure, including in Australia and Scandinavia where it has been argued that the whip should only be permitted to be used on safety grounds to avoid dangerous situations.[13] In the UK, the British Horseracing Authority opened a public consultation on the use of the whip in 2021. The Royal Society for the Prevention of Cruelty to Animals (RSPCA) responded calling for the end of the use of what it called the "unedifying spectacle of horses being repeatedly struck with the whip, particularly approaching the finishing line".[14] The RSPCA's case is based on scientific research into the effects of whipping and the likelihood that horses will feel pain, but also presents evidence to suggest that whipping increases the risk of horses falling and becoming injured.[15]

Perhaps even more controversial than whipping is the question of racing over jumps. Racing over jumps does not take place widely across the racing world and is limited mainly to north-west Europe and, until recently, parts of Australia. National hunt racing remains a significant feature of British racing, accounting for over a third of all races and for the greater proportion of racehorse deaths each year. Pressure has built up around the world against racing over jumps because of animal welfare concerns and mounting evidence of the greater risks of serious injury to horses. In Australia, political pressure has resulted in racing over jumps being increasingly outlawed. By 2009, only Victoria and South Australia still permitted jump races.[16] South Australia banned racing over jumps in 2022, leaving Victoria as the only state where it continues to be permitted.

Campaigners have also become increasingly focussed on the welfare implications of some practices in other elite equestrian sports such as dressage and showjumping. Most recently, a horse used in the modern pentathlon at the Tokyo Olympic Games in 2021 was seen and videoed being whipped and kicked by its rider, which was widely viewed on social media and prompted an international controversy. A Danish equine welfare streaming channel, Epona.tv, launched in 2007, produces video content drawing attention to animal welfare problems in equestrian sport.[17] Concerns include the use of anti-inflammatory painkillers which enable horses to continue to compete while lame by masking pain. During the 2000s, there were a series of horse doping scandals surrounding international equestrian sport. In 2004, 4.8 per cent of doping tests were positive and four horses failed their doping tests at the Athens Olympics that year.[18] Co-founder of Epona.tv, Julie Taylor, reports the cases of horses where damage to cartilage and tendons mean they are unable to jump. Through surgery, the nerve to the site of the injury can be cut, which stops pain signals being sent to the brain. From 2013, a ban on horses competing after having this procedure was lifted by the Fédération Équestre Internationale (FEI), the international governing body for Olympic equestrian sport.[19] Equine veterinary science is increasingly deployed by campaigners to show how horses that have to land from big jumps and make fast, sharp turns are more likely to injure their front legs.[20] A British survey in 2014 found that half of horses participating in equestrian sport, which were presumed sound by their owners, were diagnosed as lame.[21] A more recent study also found that 85 per cent of equestrians, including experienced professionals, fail to recognise signs of negative emotions in horses.[22]

Epona.tv has drawn attention to the practice of rollkur, a training technique used in dressage which pulls the horse's head close to its chest and so coercively limits its scope to resist the rider. The practice was banned by the FEI in 2008 as a form of abuse. A Welfare Subcommittee of the FEI Veterinary Department led on devising new rules, defining rollkur as the prolonged and excessive bending of a horse's neck, while a brief and gentle technique lasting only a few seconds would be termed 'hyperflexion' and would be permitted. Drawing the line between what was permissible and prohibited proved difficult and has become the focus of dispute between equestrian organisations and animal welfare campaigners. Prolonged hyperflexion (or rollkur) causes the horse's tongue to go blue and hang out of the mouth because of the severe pressure of the bit. In 2009, footage of blue-tongued horses at an FEI World Cup Qualifier event in Denmark was posted on YouTube. The video "spread like wildfire", prompting extensive concern about horse welfare in the sport.[23]

The conceptual basis of animal welfare centres on five 'domains' of nutrition, environment, health, behaviour and mental state.[24] The five domains have become a widely adopted tool for understanding and assessing the impact of environments and practices on animals' welfare. The framework crucially establishes mental wellbeing and emotional needs more centrally within considerations of how animals are treated. It calls into question several established practices in elite equestrian sport, so campaigners press for reform and call for equestrian sports to be dropped from the Olympics, for example. Julie Taylor's book, *I Can't Watch Anymore*, traces her experience in pressing these concerns and the response of the equestrian sports organisers. "One powerful indicator of the true status of horses within horse sport is that the animal rights movement is regarded by equestrians with terror and hostility", she argues.[25] She sets her case to the International Olympic Committee in the context of changing social values, but also mobilising the latest science on animal welfare and wellbeing.

> The younger generations, to whom you seek to impart Olympic values, are flocking to social justice issues, including non-human animal rights. They are giving up cheese because they don't want to rob a calf of its mother. They are giving up eggs because they care about the reproductive health of hens. Going to Sea World is no longer cool. How are you going to sell these young idealists the performance of horses who are being forced to participate, who live their entire lives in crates, and who are made to spend hours and hours every week trying to balance on the back of a truck with their heads tied to the wall? Horses who are spurred until they bleed, who have their mouths strapped shut, their tongues crushed, their lips torn, and whose trauma and social deprivation render them physically ill and mentally unbalanced? All for human entertainment.[26]

A 2022 YouGov survey found that 20 per cent of respondents in the UK considered that horses should no longer be used for sport at all, and 60 per cent felt that horse use in sport should be more strictly regulated.[27] Later in 2022, a public opinion survey commissioned by the Equine Ethics and Wellbeing Commission across 14 countries found that 65 per cent of respondents were concerned about the use of horses in sport, including 64 per cent of respondents in the UK.[28] There is widespread concern about how horses are treated in horse sports that the sports' governing bodies are finding it increasingly difficult to ignore.

Human Behaviour and Horse Wellbeing

Owning and riding a horse is a resource-intensive and time-consuming pastime.[29] Animal welfare organisations have concerns about people who might grudgingly find themselves with a horse they care little about and find it a burden. A range of horse sanctuaries exist across the UK to provide homes for unwanted horses. Nevertheless, most horses are owned and looked after by horse enthusiasts keen to take good care of them. Horse ownership is often an important part of the social identity and daily time commitment of horse owners. As knowledge and understanding about horse welfare develops, practice in equine care adapts, but the relationship between scientific knowledge and action in the field is not straightforward. Guidance and advice about best practice changes over time, and there is the question of how the individual behaviour of horse owners might be collectively influenced, a question not specific to horse owners which applies across areas of animal welfare and beyond. How has science helped highlight the need for changes in how horses are looked after to improve animal welfare standards? And what models of behaviour change among horse owners underpin efforts to improve horse welfare?

In 2011, the European State Studs' Association sought to capture the key concerns surrounding horse welfare in Europe.[30] They used the concept of dignity, a value that is intrinsic to an animal and that, they argued, must be respected in all interactions with it. Debasement and excessive instrumentalisation were seen as practices which can violate an animal's dignity. Their report noted marked changes in social values around the treatment of animals and called for an independent permanent commission to consider the ethical treatment of horses. A decade later, in the light of significant scientific advances in the field of animal welfare, the work was updated for the Swiss council for the equine industry, COFICHEV.

> Legal rules evolve along with the mores of a society and do not, themselves, have a moral character. Should we therefore only apply the laws of the moment and consider that anything that is not forbidden remains implicitly allowed? Or, on the contrary, are we capable of going beyond mere legality and asking ourselves: if we want to do what's right, or avoid doing what's wrong, how can we adjust our behaviour?[31]

COFICHEV's report attracted considerable interest, including in the UK horse sector.[32] It described a paradigm shift in public attitudes towards the use of the horse, threatening the sector's social licence to operate. Horse welfare gets approached from two different perspectives. The first is the ethic of responsibility, which tolerates the utilisation of animals but in exchange for the care and subsistence provided for them. This pragmatic approach accepts the asymmetry in the relationship between humans and horses, which stems from what is seen as the irreversible status of domesticated animals. A newer perspective, a more conviction-based ethics, rejects the speciesism of the asymmetrical relationship and advocates the abolition of all forms of use of living beings.[33] The report acknowledges that aspects of equine sport pose risks of physical and physiological harm to horses. Compared to recreational equestrian sport, elite competition can pose higher degrees of physical and mental strain. It suggests that "there is still significant room for improvement, and higher-strain practices are not being satisfactorily reduced" and that this is fuelling a sense of public unease with the welfare of horses secondary to the prestige and financial rewards that drive the desire to win.[34] It identified a set of ethical issues relating to the general management of horses including the management of uncastrated males, the management of castration, restrictions on movement, branding and marking, exaggerated or inadequate care, inappropriate and overly restrictive tack, hoof-care, transport, medication and end of life care.

By nature, horses lived in large territories, so restricting their movement areas in fenced paddocks is a constraint that can affect their behaviour and welfare. Different types of fencing have different welfare implications, and paddock areas below 10,000m^2 can lead to more aggressive behaviour.[35] After spending the first part of their lives in a group, young horses entering training may be managed as individuals which can cause strain and social interaction among horses is an important part of a high-welfare environment. (In Switzerland, horses are only permitted to be kept in groups of at least two.) Horses come under stress when being transported, whether by road in towed horseboxes or specialist lorries or by air, yet this is necessary for sporting competitions and for thoroughbred breeding. There are also a host of ethical issues around the use of medication for horses, including in sporting competitions. The COFICHEV report acknowledges that "ideal and absolute harmony is impossible to achieve" but argues that moral responsibility is to question practices and right wrongs.[36]

Equine veterinary professionals engage in debates about the ethical care of horses, drawing on scientific research into animal welfare and behaviour. Madeleine Campbell, an equine veterinarian and clinical specialist in equine reproduction, draws parallels between the evolution of thinking about the ethics of animal welfare and the development of feminist politics. "Increasingly, humans identify animals on an individual basis and increasingly humans simultaneously view all animals as a group within society whose interests have to be considered".[37] As new scientific knowledge emerges, what are the means by which changes in human behaviour are promoted, encouraged and legislated for? Animal welfare organisations play a key role within the horse sector in conveying the latest scientific research and its implications for horse management practices. Roly Owers of World Horse Welfare, for example, is a prominent speaker at horse sector events and has emphasised the concept of social licence to operate and its obligations to continually reflect on practice, engage with stakeholders and explain how welfare is managed in the sector.[38] In addition, institutions have developed with a mission to promote behaviour change and improve practice among those responsible for looking after horses. For example, the Equine Behaviour and Training Association (EBTA) involves a group of experienced horse owners, behaviourists and academic researchers committed to understanding equine behaviour and "working to improve the management and training of equines in the UK and to promote their physical and emotional wellbeing".[39] The organisation is focussed on equine behavioural science and its practical application to training horses. It provides support, including videos, leaflets and published resources, and conducts research projects. On its website, it states:

> We acknowledge our responsibility to the horse as a domestic species taken out of their natural environment and required to cope with a variety of ridden and management demands. Furthermore, we recognise that typically the answer to any behaviour problem lies with us, the humans, rather than the horse. Therefore, we should look to our own behaviour change so as to prevent the problem arising, be that by adapting our management practice or amending our riding and training.[40]

In addition, Human Behaviour Change for Animals (HBCA) was founded in 2016 to provide research and educational activities to apply the science of behaviour change to issues affecting animals. It argues that "to create effective positive change for animals we must consider the human animal",[41] and focus on how pro-animal welfare habitual behaviours can be developed in people. One issue they address is the problem of increasing obesity among horses in the UK. They point to research funded by the Horse Trust that has suggested that many horse owners prioritise care and grooming over exercise, with greater concern being placed on the risk of horses being underweight than overweight. HBCA promotes good practice and disseminates guidance on equine diet and weight management.[42] Owners may find it difficult to differentiate between obesity and the shape their horses are 'meant to be', constructing equine fat as an integral part of the equine body. Even when they do realise that a horse is overweight, owners can find weight management difficult because feeding a horse is so intricately bound up with their notion of care. The concept of 'responsible ownership' therefore has to be redefined, and a new moral order has to be constructed to help motivate and incentivise horse owners to comply with new norms as scientific research sheds light on not only the problem of horse obesity but also the difficulties horse owners face through the social and individual practices of horse feeding. Promotion of good practice is one strategy, through horse organisations such as World Horse Welfare, the British Horse Society, EBTA and HBCA. However, a significant proportion of horse owners and riders are not members of any horse organisations. (There are over 300,000 horse-owning households in Britain, but only 110,000 members of the British Horse Society, for example.) Peer pressure may be exerted where horses are housed and cared for in more communal settings such as liveries, for example. It has been estimated that around 60 per cent of leisure horses are kept in livery yards, so livery yard managers can be key agents in developing good practice and helping promote behaviour change among owners and riders.[43] However, changing practice across the whole population of horse owners can be challenging.

Human–Horse Interaction and Human Wellbeing

While horse behaviour may have traditionally been the preserve of animal scientists, the interaction between humans and horses is inevitably an area of knowledge that draws upon both natural and social sciences. The relatively recent academic fields of posthumanism, human–animal studies and critical animal studies have begun to grapple with the nature of human–horse interactions. Those who spend time with horses are quick to point out that horses can interact with different humans in different ways and respond differentially to human emotions and behaviours. Part of the attraction of spending time with horses is to engage with the ways they 'know' and interact. Social scientists are beginning to open the question of horses as social actors, and the dynamics of human–horse relationships are the subject of increasing attention.[44] The topic has been given new impetus by the growth of 'equine assisted therapy', the use of horses to assist in enhancing wellbeing and supporting people participating in therapeutic programmes, a practice in which relationships between horses and humans are actively and purposefully constructed.[45]

What do horses do in equine assisted therapy? One claim is that interactions with the horse help provide opportunities for the therapist to establish a productive relationship with the client. The horse is a catalyst and stimulant to a more conventional therapeutic process. A related benefit is that the presence of a horse helps a therapist establish a new dynamic with a client through the shared experience of interacting with the horse. It is argued that a different working environment, coupled with interactions with a large animal, can place the client and therapist on a more equal footing, especially if the therapist is not used to dealing with horses. Compared to the usual power relations of a conventional therapy room, both client and therapist are equally out of their comfort zones. Horses therefore contribute to a more conducive environment in which therapy takes place.[46]

Working systematically with horses to improve human health and wellbeing dates to the 1950s and was initially focussed on children and adults with developmental conditions and physical disabilities. In the UK, the Riding for the Disabled Association founded in 1969 is a federation of around 500 independent local groups that provide therapeutic horse-riding, carriage-driving and other horse-related activities. The term 'equine-assisted therapy' only began to be used in the early years of the twenty-first century, but interest in the practice has grown significantly in the past decade. Henrik Lerner and Gunilla Silfverberg have used a capability approach framework to map out the ways in which equine-assisted therapy might benefit clients and help them to live a flourishing and dignified life. Capabilities include the capability for life, bodily health and integrity. Exercises and the relation to the horse can help contribute to a better understanding of physical capability and so enhance the capacity for living a good life. Research with stroke victims has shown how equine-assisted therapy can help raise aspirations around rehabilitation. Capabilities around senses, imagination, thought and emotions can be developed through equine-assisted therapy, and it can develop emotional literacy among its participants. Because horses do not judge people as a result of their social, mental or physical status, but only as a result of how they behave towards the horse, interacting with horses provides a helpful environment in which to develop capabilities around affiliation and affinity. Working with horses can help build confidence for social interactions with other people.[47] Benefits have been reported for a variety of conditions including schizophrenia-spectrum illness, autism-spectrum disorders, attention-deficit/hyperactivity disorder, social anxiety, dyspraxia, attachment disorders and depression.[48]

Equine-assisted therapy has become a prominent feature of rehabilitation therapy among military veterans, particularly in North America. After the September 11 terrorist attacks in the United States in 2001, over 2.75 million military services personnel were deployed in combat zones including Afghanistan and Iraq. American military veterans now experience high rates of psychiatric and substance abuse disorders. Some 87 per cent of veterans report exposure to at least one potentially traumatic event, and post-traumatic stress disorder (PTSD) is reported among 23 per cent of veterans of the most recent military conflict in Iraq, for example.[49] Over recent years, there has been a growth in demand for and provision of equine-assisted therapy for military veterans in the United States. Between 2009 and 2016, the number of equine centres in the United States accredited by the Professional Association of Therapeutic Horsemanship International providing services to veterans grew more than three-fold from 89 to 335.[50]

William Marchand and colleagues identified 10 research studies into the effectiveness of equine-assisted therapy for military veterans with PTSD in the United States. Evaluation studies have begun to suggest that equine-assisted therapy shows promise as a therapeutic programme compared to other alternatives.[51] Reviewing the precise mechanisms which bring benefits, the bond between human and horse is suggested to be particularly therapeutic for military veterans. Bonding results in an affectionate and comforting relationship. One participant explained, "when you're with a horse, they give you kindness and compassion and love and they don't expect anything".[52] This is helpful where individuals have developed difficulties in forming and maintaining human-to-human bonds. In a Canadian study, participants identified several differences between equine-assisted therapy and the clinical care they had previously received. Half the participants emphasised the distinctiveness of the experiential or practical nature of equine-assisted therapy compared to conventional psychotherapy. Participants considered equine-assisted therapy to represent an alternative and distinctive way of approaching their problems and found the presence of the horse in the intervention played a key and pivotal role. The evaluation found that most participants had the firm intention to continue equine-assisted therapy sessions and recommend it to their peers.[53]

Recent research in neuroimaging is highlighting how equine-assisted therapy can affect functional and structural changes in the brain of people with PTSD, a condition estimated to affect 9 per cent of the American population and up to 30 per cent of US military veterans.[54] This is a transformation in the role of the horse from an instrument of war over many centuries to an aid in assisting human recovery from war in the twenty-first century. Therapeutic courses are being offered in the UK through a widening range of service providers, and the practice is increasingly featuring as a discipline for research and training in equine science.[55]

In the UK, equine-assisted therapies have been developed for military veterans by HorseBack UK, a charity established by Jock and Emma Hutchison. In November 2008, they moved to Aberdeenshire with a plan to establish a horse-riding business on a farm.[56] After ex-military friends visited and suggested engaging with the horses would be good for military personnel returning from war, Jock, an ex-Marine, approached a local commanding officer based in Arbroath and offered the farm as a place where injured marines could visit as a break from clinical recovery. In 2009, HorseBack UK was established as a charity and began regularly taking wounded servicemen and women and introducing them to horses. It has since diversified into providing programmes for school children and for Aberdeenshire Councils' Substance Misuse Department. The charity has formalised its provision to provide nationally accredited qualifications. Accredited by the Scottish Qualifications Authority, it provides a certificate in 'Personal Development Through Horsemanship' as a three-week, three-phase residential programme. Phase I involves engaging with other participants, setting life goals and completing team and leadership tasks, working initially with horses on the ground but then riding them. Phase 2 develops horsemanship skills and skills to increase calmness. Phase 3 involves planning, preparing for and participating in a camping expedition on horseback. The experience can be transformational for participants, and the charity's website contains powerful personal testimonies about the effects of the activities. One reads:

> I am an individual who has been suffering from PTSD for 10 years. On a regular basis I have to make the decision whether to live or die, to stay in the depths of despair, depression, worthlessness, guilt or horror. Talking with people who have been through the same thing helps me, but today I met my horse. She could feel everything that was going on inside me so there was no escape, I had to stand tall and be strong. I had to let her know how emotionally fragile I am but at the same time gain her respect and be strong to protect and lead her. We connected and I felt strong, capable and in control for the first time in many years. This has breathed new life into me as for 10 years I've been lost and now after my stay at HorseBack I am found. Thank you. Thank you. [57]

Equine-assisted therapy is still in its relatively early days of development but is arousing increasing interest among rehabilitation professionals. The military conflicts of the twenty-first century have expanded the population of military and ex-military personnel with rehabilitation needs. In the United States, the growth in the use of horses for therapeutic purposes has been marked, and there is much personal testimony that is powerful and positive. In the UK, the use of horses working with military veterans is a more recent phenomenon, but positive impacts have been reported. Evaluations are beginning to be produced to methodically assess the impacts and value of such approaches in the United States and Canada.[58] As this body of research develops, it may help support and justify the further growth and development of the practice. This new area of therapeutic and neurological research may shape the development of this relatively novel use of horses in future. Few have yet drawn the paradoxical association between the role of the horse in the development of warfare and military culture over several thousand years and this nascent role in helping to deal with the human consequences of militarism.

Horses, Nature Conservation and Climate Change

A further area of advancing scientific knowledge likely to influence the future of the horse concerns nature conservation, climate change and sustainability, which is reshaping whole swathes of economic and social life. Scientific recognition of the effects of greenhouse gas emissions on global temperatures became more widespread during the 1970s and 1980s and culminated in 1988 in the UN's establishment of the Intergovernmental Panel on Climate Change to review the evolving science and advise the world's governments. Successive international treaties since the Rio Summit in 1992 have strengthened commitments to address climate change by reducing greenhouse gas emissions, and the Paris Conference in 2015 agreed that emissions should be reduced to ensure that the rise in global temperature is kept to less than 2°C and ideally under 1.5°C. National governments are tasked with preparing plans to reduce emissions sufficient to meet this global goal. In the UK, a target of transitioning to net zero emissions by 2050 was enshrined in law in June 2019, and the government published a net zero strategy in October 2021 just before the Glasgow Climate Conference.[59]

In the UK, mitigating greenhouse gas emissions has initially involved a transition away from the most polluting fossil fuels to renewable energy sources to generate electricity. The sale of new petrol and diesel vehicles was banned from 2030, and road and rail transport systems will transition to renewable energy sources. As the high emitting sectors decarbonise, the role of food production and rural land use in emissions and in climate change mitigation is becoming increasingly prominent as a challenge in delivering the net zero transition.[60] Meeting the net zero target by 2050 is likely to require significant changes in rural land use. The Climate Change Committee estimates that at least 30,000 hectares of new forest and woodland will need to be planted each year by 2025 and planting rates will have to reach 40,000 hectares a year in the 2030s to ensure that the UK's forested area rises from 3.2 million hectares to more than 4 million. Land may also need to be planted with energy crops to contribute to demands for renewable energy. Food production levels will need to be maintained, but on less agricultural land, with the Climate Change Committee anticipating that around a third less land will be available for food production in the UK by 2050 because of other demands. The scale and implications of the climate change and net zero challenge for British farming and the countryside have prompted a clamour for a new strategic framework for rural land use to help manage the competing priorities around food production, climate change mitigation and other uses. The outbreak of war in Ukraine in 2022 only served to heighten concerns as the conflict disrupted international trade in food commodities and triggered rising input costs, contributing to steeply rising food prices and a serious squeeze on disposable incomes across Western economies and more widely. It is in this international political-economic context that the role of the horse in sustainable rural development must be considered.

The horse industry is a source of greenhouse gas emissions, although there has not been much attention paid to this source compared to other rural land uses. The world's 58.8 million horses are estimated to emit around 1.1 million tonnes of methane a year.[61] The main types of emissions from the UK horse sector are from the horses themselves, from horse manure, the production of horse feed and the travel associated with horse care, sporting competition and breeding. Horses are not ruminants, so they do not emit methane to the same degree as cattle. In the UK, total emissions of methane per animal through enteric fermentation are estimated to be 124kg per year for dairy cattle, 76kg for beef cattle and 18kg for horses. The UK government, under its reporting to the UN, calculated that horses contributed 17 kilotonnes of methane emissions in 2021.[62]

The average adult horse produces around 10 tons of manure per year, or 9,000 litres in volume. It contains nitrogen and phosphorus and is also a source of greenhouse gas emissions.[63] Straw bedding may increase the volume of manure by two to three times. Most horses are kept in small numbers and on limited acreages, so there are limits to what can be achieved through best practice manure management and disposal compared to specialist livestock farms, for example. Guidance for horse-riding establishments in 2015 included details on manure and waste management.[64] It discusses the location of muck heaps with a view to minimising nuisance to neighbours and the risks of water pollution but does not consider greenhouse gas emissions. Recent research in Finland has begun to explore the full environmental impacts of horse manure using the life-cycle assessment approach across different waste management chains.[65]

There has been some coverage of the question of climate change and environmental impacts in the popular equestrian media, but there is a lack of focussed scientific research, and measures are little more than superficial gestures around sourcing supplies locally or minimising the use of non-recyclable materials.[66] There is currently little discussion about the possible need for systematic changes in the horse industry in the UK. Reports are beginning to be produced internationally. For example, the Swedish horse sector produced an analysis of the implications of sustainable development for the horse sector in 2016.[67] It considered the positive contribution that horse riding makes to health and wellbeing and pointed out how through their grazing horses can contribute to biodiversity and biodiversity management. It examined the significance of horse feed in sustainability challenges and pointed to soybeans in horse feed as an issue to be considered further. The emissions from the machinery and agrochemicals required for cultivation of horse feed were seen as the most significant contributor of greenhouse gas emissions. The study also considered manure management and the potential for reducing emissions through changes to the energy use of stables, seeking to reduce the transport of horses by road to the minimum necessary and considering sustainability in the purchasing of clothing, equipment and accessories.

The horseracing industry has also begun to consider the implications of long-term environmental challenges for its operations, launching an initial assessment of the sport's progress in environmental sustainability in December 2021.[68] The assessment is expected to cover greenhouse gas emissions, decarbonisation and renewable energy, resource management and the potential to enhance biodiversity. The work is funded by the Racing Foundation, a charitable trust that supports the racing industry and funds work that contributes towards positive change.[69] Elite racehorses and thoroughbreds for breeding are transported around the world by air although this is not mentioned in the launch of the sustainability review. The British Equestrian Federation has also produced a strategic plan for 2020 to 2024, but it contains no discussion of impacts or commitments on the environment.[70]

It remains unclear how the commitment to net zero by 2050 will impact upon the shape and activities of the British horse sector. Increasing pressures on rural land use may reshape the horse economy. A stronger public and political commitment to domestic food security and the need for land for energy crops may alter the economics of the rural land market, and land currently under paddocks for horses might begin to shift towards food production or mitigating emissions through tree-planting. There is certainly likely to be increasing interest in the resource efficiency of key supply chains in the horse industry and a quest for reduced emissions around travel and transport. The global trade in breeding thoroughbred horses has not yet been the focus of much significant interest or pressure, but the racing industry is beginning to consider its environmental footprint. For a sporting sector that has been conservative and defensive in the face of criticisms about animal welfare or social exclusion and elitism, the challenge around environmental sustainability may meet with a similarly defensive response.

Management of grazed pastureland is one area where the horse sector could potentially contribute significantly to climate change mitigation. With the British Horse Society recommending 1 to 1.5 acres (0.6 hectares) of pasture per animal, the total acreage of pasture given to horses and ponies might be expected to be over a million acres (430,000 hectares). If racecourses and land for growing additional fodder for horses are included too, the figure could potentially be of the order of 500,000 hectares.[71] The way this land is managed has received little scientific attention to date.[72] A recent survey of over 750 respondents with horses highlighted diverse approaches to the sustainable management of equine grazing land and a commitment to improving land management for the sake of horse health and wellbeing and environmental quality. One respondent explained:

> I used to spend hours doing "horse chores" and my horses were not happy and my land was degrading every year. My land also used to look bad, but now it looks like a proper meadow with lots of wildlife/birdlife, beneficial insects/bees. Keeping my horses in this way has led to me appreciating the environment much more. I now see how horses need to be part of an ecosystem.[73]

Some horse owners are beginning to use conservation grazing (or rewilding) approaches to managing their land where grazing horses are conceptualised as part of an ecosystem and a means of enhancing biodiversity. Those pursuing conservation grazing report improvements in horse health which they attribute to the approach. Rewilding has become a prominent feature of debates about sustainable agriculture and land management in the UK. Variously called agro-ecology, regenerative agriculture or 'nature-friendly farming', it implies managing land for the sake of nature and ensuring any agricultural production practices work within ecological constraints. It often means that levels of agricultural productivity are reduced, at least in the short term, and more extensive pasture-based systems of livestock production are favoured. Similarly, with deeper conservation practices in equine pasture management, more land may be required per animal. They are likely to lead to improved soil carbon retention and thus provide carbon sequestration to help offset emissions in the transition to the UK's net zero by 2050 target. As the science of soil carbon sequestration develops in the context of increasing pressures upon rural land use, the management of equine pastureland for environmental purposes could potentially become a growing force shaping practice in the horse sector in future.

Conclusions

Today, most horses are owned by private individuals for wholly recreational purposes, although a sizable proportion of the horse economy revolves around horseracing and sport. The knowledge systems around the horse tend to map onto these two parts of the horse world – elite sporting horses and the recreational horse. The future evolution of the place of the horse in contemporary society, and the processes and practices of caring for horses, will be shaped by new knowledge about horses, their welfare and the environments in which they live, work and are looked after.

The knowledge systems that will shape the future of the horse are highly internationalised as researchers are part of global knowledge networks. They often collaborate in research projects that cross national boundaries, exchange insights through international conferences and publish findings in international journals. To understand the changing forces at play in influencing horse breeding, the practices of horse care, the politics of horse welfare and the sustainability of the horse sector, a national perspective is highly partial and limiting. The build-up of pressures for reform in the worlds of horseracing and other equestrian sports can be found in the research and equestrian politics of Australia and Europe.[74] Racing over jumps is being more widely prohibited because of animal welfare concerns in Australian states, and the Australian experience is often cited in the UK in discussions about the challenge to the sector's social licence to operate. The Australian RSPCA have pressed the case for the banning of the use of the whip in racing. Campaigners based in Denmark are marshalling international research to press their case that equestrian sports should be banned from the Olympics because animal welfare cannot be sufficiently ensured under the current regulatory regime. Research from Switzerland is setting new benchmarks in raising standards of horse welfare among recreational horses and informing debates in the UK and elsewhere. Experience of equine-assisted therapy in the United States and Canada is stimulating interest around the world and establishing new research agendas around the efficacy of therapeutic practices involving horses. And research into the contribution of the horse sector, both positively and negatively, when it comes to greenhouse gas emissions, is advancing in Australia, Scandinavia and Mexico and beginning to prompt new research questions in the UK.

These arenas for producing new knowledge draw on conventional equine veterinary science, but also the burgeoning areas of animal ethics and environmental science. Changing social values around the treatment of animals are profoundly altering the context within which equestrian organisations are having to work and posing fundamental questions about practices that have been long established in the horse world. The 'equestrian establishment' at times looks rocked by an increasingly vociferous critique. Reviews get launched to investigate controversial issues, but animal welfare groups complain of whitewashes and establishment cover-ups. The arguments about horse welfare are likely to continue to generate conflict and controversy into the future because of the fundamental differences in philosophy and ideology at their heart.

The equestrian sector's relationship with the climate change challenge is an area where there has been little scientific research to date. However, the commitment to transition to net zero emissions will come to impinge upon the horse world before too long. In the UK, the implications of the transition to net zero are only beginning to come into focus, but already, major questions arise about the future of food production, the pattern of afforestation and the competing pressures on rural land use. The afforestation targets advocated by the Climate Change Committee and enshrined in national plans for England, Scotland, Wales and Northern Ireland require almost a million hectares of additional land to be forested by 2050, and possibly more. The horse sector occupies of the order of 500,000 hectares, or the equivalent of around half of that area. Over a century on from the point of peak horse, the horse sector sits at a critical juncture. Economically challenged, subject to strident critique, and with its social licence to operate coming under pressure, the sector lacks strategic direction and suffers from an institutionally fragmented framework of representation. In this troubled context, where next, then, for the horse in Britain?

Notes

1 Raulff, U. (2018) *Farewell to the Horse: The Final Century of Our Relationship*. London: Penguin, p.315.
2 Franzen, J. (2010) *The Rise of the Horse: 55 Million Years of Evolution*. Baltimore: Johns Hopkins University Press; Kanne, K. (2022) Riding, ruling, resistance: Equestrianism and political authority in the Hungarian Bronze Age, *Cultural Anthropology* 63, 289–329.
3 Kelekna, P. (2009) *The Horse in Human History*. New York: Cambridge University Press.
4 Raulff (2018).
5 World Horse Welfare and EuroGroup4Animals (2015) *Removing the Blinkers: The Health and Welfare of European Equidae in 2015*. Snetterton: World Horse Welfare.
6 Demuijnck, G. and Fasterling, B. (2016) The social licence to operate, *Journal of Business Ethics* 136, 675–85; Gehman, J. *et al.* (2017) Social licence to operate: legitimacy by another name, *Canadian Public Administration* 60, 293–317; Duncan, E. *et al.* (2018) 'No one has ever seen … smelt … or sensed a social licence': Animal geographies and social licence to operate, *Geoforum* 96, 318–27; Douglas, J. *et al.* (2022) Social licence to operate: What can equestrian sports learn from other industries, *Animals* 12, 1987.
7 Woods, A. (2011) From cruelty to welfare: the emergence of farm animal welfare in Britain, 1964–71, *Endeavour* 36, 14–22.
8 https://www.beva.org.uk/About-BEVA
9 https://equitationscience.com/about/
10 McManus, P. *et al.* (2013) *The Global Horseracing Industry: Social, Economic, Environmental and Ethical Perspectives*. London: Routledge, pp.139–41.
11 Wade, C. *et al.* (2009) Genome sequence, comparative analysis, and population genetics of the domestic horse. *Science* 326 (5954), 865–67.
12 Campbell, M. and McNamee, M. (2021) Ethics, genetic technologies and equine sports: The prospect of regulation of a modified therapeutic use exemption policy, *Sport, Ethics and Philosophy* 15, 227–50.
13 McManus *et al.* (2013), p.153; Duncan *et al.* (2018).
14 Kennedy, M. (2022) It's time for horse racing to wave goodbye to the whip, RSPCA Blog, 8th April 2022, https://www.rspca.org.uk/-/blog-time-for-horse-racing-to-wave-goodbye-to-the-whip
15 Kennedy, M. (2021) An end to 'encouragement'- it's time for horseracing to follow the science about whipping, *RSPCA Blog*, https://www.rspca.org.uk/-/blog-time-for-horseracing-to-follow-science-about-whipping
16 For an account of the debates in Australia about racing over jumps see McManus *et al.* (2013), pp.185–202.
17 https://www.epona.tv/
18 Taylor, J. (2022) *'I Can't Watch Anymore': The Case for Dropping Equestrian from the Olympic Games*. Copenhagen: Epona Publishing, pp.43–4.
19 See Taylor (2022), p.65.
20 Murray, R. *et al.* (2006) Association of type of sport and performance level with anatomical site of orthopaedic injury diagnosis, *Equine Exercise Physiology* 7, *Equine Veterinary Journal Supplement* 36; see also Taylor (2022), p.72.
21 Bankes, C. (2014) Half of all sport horse are lame, study finds, *Horse & Hound* 28 April 2014.

22 Bell, C. *et al.* (2019) Improving the recognition of equine affective states, *Animals* 9, 1124.
23 Taylor (2022), p.99.
24 Mellor, D. and Reid, C. (1994) Concepts of animal well-being and predicting the impact of procedures on experimental animals. pp.3–18 in *Improving the Well-Being of Animals in the Research Environment*. Glen Osmond, SA, Australia: Australian and New Zealand Council for the Care of Animals in Research and Teaching (ANZCCART).
25 Taylor (2022), p.27.
26 Taylor (2022), p.152.
27 World Horse Welfare (2022) Sector leaders discuss involvement of horses in sport, *Press Release* 21st June 2022. The survey was repeated in 2023 and found broadly similar results.
28 Equine Ethics and Wellbeing Commission (2022) *Public Attitudes on the Use of Horses in Sport: Survey Report*.
29 Dashper, K. *et al.* (2020) 'Do horses cause divorces?' Autoethnographic insights on family, relationships and resource-intensive leisure, *Annals of Leisure Research* 23, 304–21.
30 Poncet, P.A. *et al.* (2011) Considerations on Ethics and the Horse - Ethical Input for Ensuring Better Protection of the Dignity and Wellbeing of Horses, in European States Studs Association (eds.) *Heritage Symposium of the European State Studs Association*, Lipica National Stud, October 13th, 2011, p.64.
31 Poncet, P.A. *et al.* (2022) *Ethical Reflections on the Dignity and Welfare of Horses and Other Equids – Pathways to Enhanced Protection*. Summary report. Bern: Swiss Horse Industry Council and Administration, p.9.
32 Jones, E. (2022) New spotlight falls on ethics surrounding the use of horses, *Horse & Hound* 26 May, p.4.
33 See also Campbell, M. (2019) *Animals, Ethics and Us: A Veterinary's View of Human-Animal Interactions*. Sheffield: 5m Books.
34 Poncet *et al.* (2022), p.15.
35 Poncet *et al.* (2022), p.17.
36 Jones (2022).
37 Campbell (2019), p.119.
38 Douglas *et al.* (2022).
39 http://www.ebta.co.uk/index.html
40 http://www.ebta.co.uk/about.html See also Rogers, S. (ed.) (2018) *Equine Behaviour in Mind: Applying Behavioural Science to the Way We Keep, Work and Care for Horses*. Sheffield: 5m Publishing.
41 https://www.hbcforanimals.com/
42 Furtado, T. *et al.* (2018) *When the Grass is Greener: The Equine Weight Management Guide for Every Horse, Every Yard, and Every Owner*. Princes Risborough: The Horse Trust and University of Liverpool; Furtado, T. (2019) *Exploring the Recognition and Management of Obesity in Horses through Qualitative Research*. PhD Thesis. University of Liverpool; Furtado, T. *et al.* (2021) Exploring horse owners' understanding of obese body condition and weight management in UK leisure horses, *Equine Veterinary Journal* 53, 752–62.
43 Furtado, T. *et al.* (2021) Equine management in UK livery yards during the COVID-19 pandemic — "As long as the horses are happy, we can work out the rest later", *Animals* 11, 1416.
44 Birke, L. and Thompson, K. (2018) *(Un)Stable Relations: Horses, Humans and Social Agency*. London: Routledge; see also Jones McVey, R. (2018) *Reasonable Creatures: British Equestrianism and Epistemological Responsibility in Late Modernity*. University of Cambridge PhD Thesis.

45 Andersson, P. (2019) Who is the horse? Horse assisted psychotherapy as a possibility for understanding horses, pp.45–56 and Lerner, H. and Silfverberg, G. (2019) Martha Nussbaum's capability approach and equine assisted therapy: an analysis for both humans and horses, pp.57–68 in J. Bornemark *et al.* (eds.) *Equine Cultures in Transition: Ethical Questions*. London: Routledge.
46 Andersson (2019).
47 Lerner and Silfverberg (2019).
48 Marchand, W. *et al.* (2021) Equine-assisted activities and therapies for veterans with posttraumatic stress disorder: Current state, challenges and future directions, *Chronic Stress* 5, 1–11.
49 See, for example, Marchand *et al.* (2021).
50 Kinney, A. *et al.* (2019) Equine-assisted interventions for veterans with service-related health conditions: a systematic mapping review. *Military Medical Research* 6(1), 28.
51 Kinney *et al.* (2019); see also Monroe, M. *et al.* (2019) Effects of an equine-assisted therapy programme for military veterans with self-reported PTSD, *Society and Animals 29*, 577–90.
52 Quoted in Marchand *et al.* (2021), p.5.
53 Blackburn, D. (2021) Equine-Assisted Therapy as complementary practice in the treatment of operational stress injuries of Canadian Armed forces veterans, *SunText Review of Neuroscience & Psychology* 2(1), 125.
54 Zhu, X. *et al.* (2021) Neural changes following equine-assisted therapy for posttraumatic stress disorder: A longitudinal multimodal imaging study. *Human Brain Mapping* 42, 1930–39.
55 Hartpury University (2020) Research reinforces benefits of equine-assisted therapy for children, *Press Release* 17 August 2020, https://www.hartpury.ac.uk/news/2020/08/research-reinforces-benefits-of-equine-assisted-therapy-for-children/
56 https://horseback.org.uk/about/introduction-to-horseback-uk/
57 https://horseback.org.uk/about/testimonials/
58 Blackburn (2021).
59 Ward, N. (2023) *Net Zero, Food and Farming: Climate Change and the UK Agri-Food System*. London: Routledge; UK Government (2021) *Net Zero Strategy: Build Back Greener*. London: Stationary Office.
60 Climate Change Committee (2020) *Land Use: Policies for a Net Zero UK*. London: Climate Change Committee; see also Ward (2023).
61 Elghandour, E. *et al.* (2019) Equine contribution in methane emission and its mitigation strategies, *Journal of Equine Veterinary Science* 72, 56–63.
62 UK Government (2023) *UK Greenhouse Gas Inventory, 1990–2021*. Annexes. London: Department for Energy Security and Net Zero. Table A 3.3.2, p.809. https://unfccc.int/documents/627789
63 Westendorf, M. *et al.* (2019) Generation and management of manure from horses and other equids, pp.145–63 in H. Waldrip *et al.* (eds.) *Animal Manure: Production, Characteristics, Environmental Concerns and Management*. Madison: John Wiley & Sons.
64 Sinclair-Williams, K. and Sinclair-Williams, M. (2015) *Health and Safety in Horse Riding Establishments and Livery Yards*. London: Chartered Institute of Environmental Health.
65 Havukainen, J. *et al.* (2020) Environmental impacts of manure management based on life cycle assessment approach, *Journal of Cleaner Production* 264, 121576.
66 Turner, R. (2022) How all equestrians can help make the world a greener place, *Horse & Hound* 4 January.
67 Blomberg, J. and Välimaa, C. (2016) *Horses and Sustainable Development: How Can the Horse Become the Caretaker of the Planet?* Stockholm: Swedish Horse Industry Foundation.

68 British Horseracing Authority (2021) Environmental sustainability project to review racing's progress and support long-term planning, *British Horseracing Authority Press Release* 7 December; British Horseracing Authority (2022) Racing's sustainable future – help map our environmental activity, *British Horseracing Authority Press Release* 8 February.
69 https://www.racingfoundation.co.uk/about
70 British Equestrian Federation (2020) *Enriching the Lives of People and Their Horses Together. Strategy 2020–2024*. Kenilworth: British Equestrian.
71 This is admittedly a crude and simplistic estimate. Taking an alternative approach, Mark Wentein, Chairman of the European Horse Network, estimates that Europe's 7 million equines (including donkeys and cross-breeds) take up around 2.6 million hectares of land. (https://www.worldhorsewelfare.org/about-us/our-organisation/our-conference). The UK's proportion of the European population of equines would suggest it takes up around 350,000 hectares of land. The UK has more land under racecourses than many other European countries, however. It can be estimated that the land under horses may be in the range of 350,000 to 500,000 hectares. As horse numbers declined in the UK since 2010, so presumably has the land area required to accommodate and feed them.
72 See Furtado, T. *et al.* (2022) An exploration of environmentally sustainable practices associated with alternative grazing management system use for horses, ponies, donkeys and mules in the UK, *Animals* 12, 151.
73 Quoted in Furtado *et al.* (2022).
74 Duncan *et al.* (2018).

9 More-Than-Human Geography and Equine Futures

Introduction

This study has taken Britain as a national case study to trace the ways in which the human–horse relationship has shaped economy, society and culture – the feel of a nation and its more-than-human geography. In agriculture, industry and transport, the horse has played a crucial role. For more than 2000 years, horses became ever more extensively intertwined with human society in Britain. Whole swathes of urban areas were designed around them, and their uses accounted for the appearance of much of the rural landscape too. At the point of peak horse in the early twentieth century, there was one horse for every 13.7 people in Britain. The legacies of this extensive and intricate equine–human entanglement endure to this day, even if the total number of horses fell dramatically in the twentieth century.

The history of horses is also a history of power. This is not only the power to carry people and haul wagons embodied in the horse itself, but also the power that ownership and control of horses conferred upon individuals, families and military forces. Horses brought power and prestige, and a sense of British nobility co-evolved between elite society and its elite horses where horses and horsemanship embodied the performance of power. There is also the power of horses, realised through the human–horse relationship, to shape places, and this is the aspect of horsepower, what we might call 'geographical horsepower', that has remained underexplored in research and scholarship. We have examined the role of the horse in co-producing the historical geography of a nation. This concluding chapter reflects on the changing nature of the human–horse relationship. It positions the study in the context of recent work in human–animal studies and animal geographies and considers the shifts in practice and meaning, in the British context, from horses as principally agents of work and war, to creatures of sport and leisure. It considers the role of the horse in the formation of places within the UK and in the formation of the nation as a whole. It also reflects on the contemporary juncture and the pattern of forces likely to influence the future of the horse.

DOI: 10.4324/9781003454359-9

Human–Animal Studies and More-Than-Human Geography

Human–animal studies have helped conceptually unsettle our conventional categories and opened new areas of scholarship in philosophy, ethics, sociology and geography. Such studies routinely deal with what Bruno Latour calls one of the 'Great Divides', between society and nature, and between the human and the non-human.[1] They also underpin new social and political movements that materially shape the worlds of non-human animals. In her 2008 book, *When Species Meet*, Donna Haraway writes "animals are everywhere full partners in worlding, in becoming with".[2] She explains the purpose of her book as to strive "to build attachment sites and tie sticky knots to bind intra-acting critters, including people, together in the kinds of response and regard that change the subject – and the object".[3] Studying our webs of relationships with non-human animals can provide potent insights into our own development and change the ways we see ourselves.

Among disciplines involved in human–animal studies, engagements with the horse have varied. In political science, treatment of the horse is rare, with Christopher Hill's study of the politics of horseracing being one exception.[4] Leisure, tourism and sports management contain a few specialist works that have considered horse-based sport, recreation and tourism, but these are usually applied policy studies that do not engage with the conceptual themes of human–animal studies. It is in the humanities where human–animal scholars have engaged most extensively with the horse and where conventional thinking and categories have been most challenged. Margaret Derry's study of horse breeding and marketing in the nineteenth century is one early example.[5] David Anthony's study of how Bronze Age riders from the Eurasian steppes affected the development of the modern world is a landmark in highlighting the human–horse relationship in the formative history of early human civilisations.[6] Pita Kelenka builds on Anthony's work to show how horse power was both constructive and destructive, tracing the importance of the horse in the rise of the Hittite, Achaemenid, Chinese, Greco-Roman, Arab, Mongol and Turkish states of the ancient world.[7] Peter Mitchell has considered the impact of the introduction of horses on indigenous societies around the world since 1492.[8] These works have provided a foundation for an explosion of interest in the horse in history and anthropology over recent years.

Linda Birke and Kirrilly Thompson's study of the horse and social theory explores the extent to which horses can be considered social actors, which act and have effects on the social world.[9] Birke's background is in biological sciences and ethology, but her work evolved to focus on feminist studies and human–animal studies. Thompson describes herself as a 'galloping cultural anthropologist'. Their interdisciplinary study considers how interspecies relationships work and how humans and horses build social lives together. This focus on the micro social theory of human–animal relations has profound implications. It is complemented by recent historical analyses of the ways the horse features in human culture. Ulrich Raulff's *Farewell to the Horse* takes as its starting point a carving up of history into the pre-horse age, the horse age and the post-horse age, and then focuses on the "long shadow of the era of the horse".[10] In more recent volumes, Kristen Guest and Monica Mattfeld have brought together historical studies of horse breeding[11] and of equestrian cultures more broadly.[12] In *Equestrian Cultures,* they argue that equestrian histories have tended to focus heavily on the pre-modern period, but that the relationships between humans and horses have been shaped by and have shaped modernity. They are contributing to the scholarly project that Donna Landry describes as "rectifying the imbalance between the equestrian saturation of early modern culture and today's marginalisation of matters equine".[13] A key tenet of modernity has been the idea that the world can be transformed by human intervention, and the modern period has seen a shift from an idea of society as heavily dependent upon animals to one where they are objectified and compartmentalised as domestic companions or 'wildlife' or as food. Recent histories of horses and human–horse relationships are creatively unsettling the metanarratives around modernity.

Given this conceptual and intellectual vibrancy around the horse within human–animal studies in the humanities, what of geography? The questions that prompted the writing of this book centre on the material spatialities and political economy of the horse. What is the nature of the horse economy and how is it changing? Who is responsible and speaks for the horse and what are their preoccupations? What does an historical geography of the horse look like? Animal geographies may be replete with cats, dogs and all sorts of exotic species, yet this sub-field of geography presents an equine paradox – the paradox of the hidden horse. Our relationship with the horse has had a more significant material impact on our world, our more-than-human geography, than any other species, yet it features little in animal geographies. So, the question becomes not what are the reasons to study the horse, but what explains why this most influential and important of animals and its changing relationship with humanity over the ages has not attracted the scholarly focus we might expect?

One possible explanation lies in animal geographies' origins in poststructuralist and postcolonial theories. The impulse to render visible and give voice to marginalised others leads to a quest to find more and more others to give voice to. After poverty, gender, race and ethnicity, and the queer turn, a need for other others may have fed the 'animal turn', with its mission to 'bring animals back in'. However, at first glance, horses looked like animals of the elite, the sport of kings and the preserve of at least those wealthy enough to have access to land and stables. If critical post-structuralist scholarship is about helping tackle injustice, there seemed very limited emancipatory potential from equestrian inquiry. In any case, by the mid to late 1990s in Britain, horse organisations were becoming enrolled into campaigns about protecting hunting with hounds. They were on a different side of an argument about animal welfare. Equestrianism may not have resonated with the welfare sensibilities of animal geographers.

An alternative explanation might lay in contemporary social science's fetishisation of the exotic and obscure over the mundane and everyday. Studies in animal geographies have tended to focus on wild animals and conservation, in tourism and adventure, and in zoos. The importance of farmed animals in the food system has prompted work in understanding the socio-technical systems that have evolved around chickens and cows, but the horse's ambiguous position as 'not quite' an agricultural animal has side-lined it from the more-than-human geographies of the agri-food system.[14] Postcolonial studies of human–animal relations ought to have great potential for studying the horse,[15] but to date the preoccupations have been around decolonising hegemonic academic thought including Western ideas of how humans should relate to nature.[16]

A recent development with some promise of better accommodating the horse within animal geographies is interest in infrastructure and non-human life. Infrastructures enable the flow of materials and at the same time produce non-human mobilities and immobilities that can affect the dynamics of everyday life. Maan Barua writes of how infrastructures themselves can become a medium of life and non-human life itself can be cast as infrastructure.[17] The role of non-human animals can affect the functioning and effects of infrastructure, and the built environments infrastructures produce. What Barua calls "these more-than-human enfleshments and enmeshments" can help unsettle what is the largely anthropocentric scholarship on infrastructure. He shows how infrastructures can influence the circulation and spatial practices of wild animals as they use or are impeded by roads, for example. Infrastructure can also become a medium, a non-human habitus. Ships become habitats for globally mobile rats and limpets, and drywood termites turn pilings, bridges and buildings into infrastructures of their own.

There can be few animal species more heavily implicated in the development of human infrastructure than the horse. A contribution of this study is therefore to use the case of Britain to demonstrate the pervading effects of horses on the form of the contemporary city, the rural landscape, and on the national settlement structure and geography of interconnectedness forged in the pre-modern and early modern period. Horses and the equestrian infrastructure that is their legacy have shaped our world in important ways that have become lost to the contemporary gaze. As animals, they continue to make a significant mark on the peri-urban fringe around towns and cities as well as accounting for a significant portion of current rural land use. In thinking about the relationships between humans and non-human animals, infrastructure can be a useful prism, and the story of humans and the horse is one of extensive and profound co-creation of infrastructure, with a legacy ripe for further exploration.

Histories and Geographies of the Horse

Human–animal studies are considering the horse and culture, power and technological change. Kristen Guest and Monica Mattfeld explain how the modern era was associated with objectifications of animals and non-human life. Animals were not generally seen as individual sentient beings, but as species, foodstuffs, wildlife. They point out how, in this respect, "modernity's metanarrative" is destabilised when the species under consideration is the horse, probably the most influential animal in human history. Horses were creatures of modernisation and early modernity, "circulated, exchanged and physically reshaped in the service of human society and culture".[18] The strong dray horse with its shiny tack was the epitome of the industrial modernity of its day. Yet more than this, horses provide a vehicle for humanity's attempts to understand modernity's effects. As they become less materially significant, they become ever richer in meaning and symbolism. Ulrich Raulff puts it thus:

> Just as the old, solid world of horses, carriages and cavalrymen begins to crumble under the pressure of an increasingly mechanised civilisation, horses take on a more imaginary and chimeric form: they are reduced to an existence as the ghosts of modernity, and the more they forfeit their worldly presence, the more they haunt the minds of a humanity that has turned away from them.[19]

Human dealings with horses and horse breeding, with their concepts of breed, bloodline and pedigree, interplayed with ideas about human aristocratic lineage. Horses served as objects denoting status, and were exchanged as gifts between monarchs, ambassadors and nobility, but at the same time were themselves ascribed with notions of nobility among animals. "The horse is thus a complex, often conflicted, figure in human society, one that – as a result of its perceived status as at once an animal object and a humanised subject – blurs the line between human and animal".[20]

Some recent histories of the horse take a global perspective,[21] but are complemented by more situated regional and national histories of the evolution of human–horse relations in particular places and times. Ann Norton Greene examines the draft horses of the Northeast and Midwest United States and their role in nineteenth-century industrialisation.[22] Clay McShane and Joel Tarr focus on the horse's role in the growth of the nineteenth-century American city.[23] Tom Almeroth-Williams investigates how horses shaped Georgian London.[24] It is uncommon for the new horse histories produced in the last 15 years or so to take the national scale as the geographical focus. Sandra Swart's studies of the history of the horse in southern Africa come close,[25] and Yashaswini Chandra has recently produced a study of the political symbolism of the horse in India.[26] Histories of horses in particular times and places help contribute the building blocks of what Ulrich Raulff calls a "total history" of the horse.[27] For the geographer, the interest lies in the spatialities of the human–horse relationship and its evolution over time. How did humanity's relationships with horses take different forms in different places and how did the use of horses help tie places together and force places apart?

One prominent geographical conundrum is the question of the use of the horse as a source of food for humans. This is the original form of the human–horse relationship as humans first encountered horses as prey for food. Today, eating horse flesh remains matter of fact in some places while deeply taboo in others. Consuming horsemeat flared up as an issue of revulsion and great public concern in Britain and Europe in 2013. Testing in Ireland had found traces of horse and pig meat in a range of beef products on sale in several supermarket chains. The Irish Food Standards Agency found that horsemeat accounted for approximately 29 per cent of the meat content of Tesco's value burgers. Follow-up testing in the UK found that Findus lasagne contained more than 60 per cent horsemeat. The horsemeat scandal prompted a public and political furore, with Parliamentary inquiries rapidly established to get to the bottom of what was going on.[28]

Meat is increasingly at the centre of human dilemmas and controversy. "Should we eat meat?" was once a question of concern only to moral philosophers, but now it fractures modern life and marks it with its signs and symbols from the "V" signs on restaurant menus to the growing meat-free sections of supermarket aisles.[29] When it comes to humanity's defining challenge of these times, climate change, the models and charts that map our transition to a world of net zero greenhouse gas emissions over the next three decades are riddled with assumptions about reductions in meat consumption. Eating meat in the 2020s is coming to be subject to increasingly vociferous critique. From the perspective of a more-than-human geography of the horse, it is a curiosity that horsemeat is eaten in some places and not in others. It tends to be taboo and avoided in Anglophone nations, while it is eaten in continental Europe. There is not a settled explanation commanding strong consensus as to why.[30]

Two broad schools of thought address the geography of hippophagy – the act or practice of feeding on horseflesh. First is a cultural materialist perspective in which culture reflects economic circumstances. The British command of international trade and access to plentiful supplies of beef from the new world left eating horse meat unnecessary. The British turned from eating horsemeat because they enjoyed plentiful supplies of alternative meats and so were able to afford to be disgusted by the very thought. A second approach is more strongly cultural. The book of Leviticus in the Bible stipulates that only cloven hooved animals are to be eaten, and Pope Gregory III banned eating horsemeat in 732. These are thought to be the source of the idea that eating horsemeat was somehow un-Godly or uncivilised. By the early modern period, the horse in Britain was coming to be increasingly seen as a 'noble' animal which reinforced the taboo. Horsemeat became cast as "a disgusting foodstuff redolent of barbarism and deprivation".[31] However, horsemeat consumption re-emerged in the nineteenth century across western Europe. Its sale was legalised across many states and regions including Denmark (1807), Bavaria (1842), Baden (1846), Saxony (1847), Austria (1847), Belgium (1847), Switzerland (1843), Prussia (1853), Norway (1855) and Sweden (1855). By the end of the nineteenth century, consumption of horsemeat in large German cities was reported to be almost as great as that of beef or mutton.[32]

In France, horsemeat was legalised in 1866 and its consumption was promoted by elite scientists, doctors and veterinarians. Dedicated horsemeat butcher shops were opened, and horsemeat consumption became adopted by a significant proportion of the working-class population. Indeed, not eating horsemeat came to be seen as wasteful. By 1929, almost 75,000 horses were being sold in France for human consumption, and during the 1930s consumption increased among the French bourgeoisie, although it was largely an urban practice and less common in rural areas.[33] In the UK, the sale of horsemeat had not been illegal, as it had in early nineteenth-century France, and it was sold openly as food for cats and dogs. It did get mixed with other meats for human consumption and made its way into adulterated sausages, for example. Dead or near-dead horses were handled by London's knacker's yards that killed 26,000 horses a year by the late nineteenth century and produced glue, leather and pet food from the process. This trade was recognised and regulated. However, the British urban working class remained much more averse to eating horsemeat than their French counterparts. Despite efforts to promote the idea, British horsemeat consumption was hampered "by a combination of sentiment, gustatory conservatism, bad cooking, working-class scepticism, agricultural economics, uncooperative butchers and the lack of strong public support from scientists".[34] The debate about eating horsemeat became spikily nationalistic. "John Bull is resolved to eat, drink and do only what he has become accustomed to. He wants none of your foreign kickshaws, frogs, and snails in fricassees, or sea slug, or bird's nest soup, or horse flesh steak", wrote Peter Lund Simmonds of British culinary reactionaries.[35] Horsemeat stores were opened in Britain, but primarily to cater for continental refugees and immigrants. Although the British did not develop a taste for eating horses, human consumption did become the fate of many British horses. By the early twentieth century, over 50,000 old horses were transported to Belgium and the Netherlands each year for slaughter for meat, and a century later the UK was still exporting almost 2,200 tonnes of horsemeat a year.[36]

The story of the nineteenth-century turn to horsemeat consumption illustrates not only how geography matters in any total history of the horse, but how human's roles and relationships with horses – the uses, functions and flows of horse bodies, and the meanings and values ascribed to horses – were intricately bound up in wider processes of social, economic and technological change. Geography shaped the human–horse relationship, and vice versa. Just as horses helped forge the industrial revolution, the building of empires and the forging of nations, the variable uses and meanings of horses helped constitute geographical differences in the equestrian sectors of different places. Of the almost 60 million horses estimated to live with us on Earth, the vast majority remain working horses. Then, there are the populations of feral horses scattered around the world, a legacy of past domestications and releases. There is a geography to the world of horseracing and to Olympic equestrianism, as well as to the popularity of keeping a horse for simple recreational purposes. Horses have different uses and meanings in different places.

The world of horseracing has its international hubs in Britain, France, the United States and Australia. International equestrian sport also has a marked international geography. It tends to be dominated by the most affluent nations of the world – Germany, France, Britain, Scandinavia. The international governing body for equestrian sport, the Fédération Équestre Internationale (FEI), is embarrassed by the socio-economic and geographical patterning of participation which undermines the Olympic value of universality. It seeks to project globality and is keen to point out that it has a membership of 136 affiliated national federations from all around the world. However, real participation and engagement is not so extensively spread. The Ethiopia Equestrian Association has been affiliated with the FEI since 1997 but as of July 2021 had never registered a single FEI athlete or horse. Over a third of affiliated national equestrian federations listed by the FEI had ten or fewer registered athletes in 2019.[37] At an international meeting in Rotterdam in 2019, the President of the German Equestrian Federation, one of the world's biggest and oldest, remarked that only about 20 per cent of FEI-affiliated national equestrian federations could really be classed as "riding countries".[38] The others make up the numbers and help to project the impression of a broader geography of participation. Julie Taylor of Epona.tv has revealed how in 2019 there were 23,300 European FEI athletes registered to compete in the three Olympic equestrian disciplines, almost 70 per cent of the total, but considered even this to be an under-representation of European dominance.[39] In this example of international equestrian political geography, she points to what she calls a "contrived universality" and argues that it is undermining the legitimacy of equestrianism as an Olympic sport and also undermining its standards. Globally, the horse and horse sport have a distinctive and patterned geography, and this continues to shape and influence the affairs of the horse world.

Whither the Horse?

The more-than-human geography of Britain's equine world and its legacies provides a case study of the evolution of a national economy, society and culture through the relationships between humans and horses. It is set of social, biophysical and infrastructural relationships that change over time. The case study provides a geographical focus, but the co-evolution of the horse world and the geography of Britain is not a bounded matter. Britain's more-than-human geography of the horse is also constituted by the flows of horses, horse people, ideas and accoutrements to and from beyond British shores. This geography of international equine interconnectedness affects the evolution of horse sectors around the world. After domestication, the horse had become an increasingly significant partner in the development of human civilisations, states and military forces. Their growing use as a vehicle for, and instrument of, military conflict had been an important determinant in the rise and fall of empires and the ebb and flow of migration and conquest initially across central Eurasia and the Middle East, but increasingly spreading more widely to cover the whole of the Eurasian landmass and North Africa.[40] Horses played their part in the development of the Roman Empire and in the rebellions against it and, in Britain, became established as a means of hunting quarry for food and sport among kings and nobility in the centuries after the fall of Rome. By the time of the Norman conquest in 1066, horses were a common feature of warfare, pageantry and everyday life. Over the next couple of centuries, whole swathes of land, up to a third of England's territory, became subject to special forest laws designed with the specific purpose of protecting the sport of hunting on horseback.

Through the early modern period, horses became increasingly important in agriculture and transport and a sophisticated sector developed with its own geography of flows of people, horses, information and money. At the same time, the powers of the horse became more effectively harnessed in the development of a national communications infrastructure that helped integrate England, then eventually the UK, as a unified nation. The central role of horses in territorial communication and postal services for three centuries helped forge a sense of a national population in touch with one another. As clusters of inns, stables, farriers and other businesses built up around coaching stops, horse travel shaped our national settlement structure along Britain's main strategic routeways and the growth of market towns. The use of horses in gin mills helped power manufacturing and processing in the years before the industrial revolution. They carried people and hauled freight. They were our principal source of power and rapidly extended mobility over land. They were a central feature of imperial domination, a key instrument of empire. They bore the managers who supervised the work of slaves on sugar plantations and were used to chase down runaways. The city of the eighteenth and nineteenth centuries was increasingly moulded around the horse, which determined the width of streets, the need for stables and an infrastructure to manage what went in and came out of each end of the horse. Accommodating stables and coach-houses in mews was integral to the development of large swaths of London's urban form. Even with the advent of the railways, more and more horses were required to ferry people and goods within towns and cities. Indeed, it has been argued that the major development in urban transport in the Victorian Age "was not the introduction of mechanical traction but the greater supplementation of human by animal power".[41]

From the early twentieth century, the transformation in the role and place of the horse has been dramatic. The moment of peak horse cannot be accurately pinpointed because of the limitations of data gathering on horse numbers and will have varied around the world. In Britain, it was likely between 1901 and 1911.[42] In the United States, it was around 1915[43] In South Africa it was the early 1920s.[44] In Finland, horse numbers did not peak until the 1950s.[45] The growing use of the machine gun during the First World War and the development of electric-powered trams rapidly killed off the use of horses for war and urban transport. Here their numbers dropped quickly. The London General Omnibus Company used 16,714 horses in 1901 but ran its last horse-drawn bus just ten years later.[46] Their decline in agriculture and some areas of haulage was less steep over the decades from the 1920s to 1950s, and pit ponies continued to be used in British coal mines through much of the twentieth century.[47]

Urban de-horsification was experienced as a burst of change in the early twentieth century, followed by a longer and more drawn-out rural retreat. Historians looking back on this period have been much more inclined to chart the spectacular uptake of the new technologies rather than the rendering redundant of the old.[48] The collapse in horse numbers would have made a major impact on the prospects for horse-dependent businesses and employment, but this does not register strongly in the social and economic commentary of the time. This major economic change occurred several decades before the deindustrialisation of heavy industries in Britain that was to prompt the development of state interventions to facilitate adjustment and ease the pain of rapid sectoral extinction. There were no Structural Funds or regional aid to smooth the collapse of the horse economy. The end of the age of the horse may have cast a long shadow over the first half of the twentieth century, but it was nevertheless a seismic shift in the transport of people and goods and in urban infrastructure. It brought transformational change in the material world of work and travel, but also the atmospherics of everyday life in towns and later villages. The end of the age of the horse also brought a set of profound shifts in meaning around the horse and its place in economy and society.

The first shift was from production to consumption. Although horse-racing had existed in Britain for centuries, until mid-twentieth century most horses were working horses hauling, carting and carrying. The horses that remained working in the mining industry throughout much of the twentieth century were not generally visible. The working horse largely faded away, leaving the dominant equestrian figure of the recreational horse (including sporting horses). Associated with this shift in function was a shift in geography. Through the nineteenth century, the horse population had become increasingly concentrated in towns and cities, even more spatially concentrated than the human population. By the 1920s, most horses were in the countryside beyond towns and cities. The electric tram and motor engine purified the city of the horse, which became a pastoralised animal of the fields and farms once more. A third shift was in the bodily form of the horses themselves. No longer required to haul wagons and omnibuses, horses were less likely to be bred for strength and draught power. Speed, agility and other aesthetics replaced strength, and the larger draught horses declined. By the 1980s, only a few thousand could be found and after that they became rarer still.[49] Finally, the social complexion of the people interacting with horses changed during the twentieth century. The horse world became more feminised and, although there remained pockets of working-class horse use and ownership, horses gradually became the more exclusive preserve of the more affluent middle classes who could afford to keep them.

Horses became more strongly associated with a particular stratum of British society – the horse set. A dedicated equestrian media helped reinforce a sense of a single, insular community centred on leisure and sporting practices. *Horse & Hound*, with a circulation of over 40,000 and over 130,000 followers on Twitter, is one of a small number of magazines devoted to all things horse. External perceptions of the horse world are shaped by popular culture including, for example, the romantic equestrian novels of Jilly Cooper or images of the royals at play. However, the dominant idea of a horse set, actively promoted by equestrian media and organisations, does conceal a diversity of communities and social groups still involved with the horse. Horse charities and social enterprises continue to operate in deprived inner-city areas pursuing a social mission.[50] Gypsy and traveller communities still congregate at the annual Appleby Horse Fair in Cumbria to display and trade in horses.

Three million riders and 6 million racegoers help support an equine economy of around £8 billion turnover a year in Britain. And as the sporting and recreational uses of horses have become more diversified and specialised over the past half century, the range of organisations representing different slithers of the horse world has become greater. In a radical turnaround from their role in the cavalry charges of imperial wars, horses are now used for therapeutic purposes to treat military veterans with post-traumatic stress disorder. This is but one strand of a wider movement of equine-assisted therapy, but it is an area where interest in and use of horses is growing. The human–horse relationship in the twenty-first century is a confused one. Over a hundred years on from the point of peak horse in Britain, old dominant meanings and narratives have faded away or been undermined. There has been an economic transformation from a dependence on the oat-powered engine to the petrol-powered engine, and from a creature of work to one principally of sport and leisure. New social forces gather purchase and influence, and loom around the world of the horse. Concerns about animal welfare rattle around within equestrian organisations who grapple defensively with how to handle this new 'problem'. The economic crisis since Brexit, Covid and the Ukraine war also poses challenges for the horse economy. The gambling world shifts towards online betting on football, thus threatening the income streams that flow to horse-racing. Climate change and the transition to net zero may bring new challenges for those who own, care for and manage land for horses. The horse economy and society in Britain sits poised at a crossroads, and without a clear sense of direction. Horse welfare bodies point to the concept of the social license to operate as the way to think about the future, and the place of horse sport and horse-riding within it.

Shaping the Future of the Horse

In 2011, Peter Edwards, author of invaluable histories of the horse in the early modern period, and his colleagues pointed out the paradox that it is "the taken-for-granted centrality of horses to human lives in the past" that has rendered them almost invisible to history.[51] Histories of the horse have become much more common over the past decade. There is a sense of this body of scholarship gathering pace, and horses are becoming rendered more visible in history. Geographies of the horse lag far behind. While the more-than-human has been embraced by geographers and there is much work on viruses, disease and the implications of microbiology for social science and social theory,[52] the most mundane, common-place and ubiquitous of animals, the horse, has not excited much interest. A century past the point of peak horse is a moment to reflect on the horse-drawn trajectories that shaped our towns, cities and countryside, and the legacies of the age of the horse for the more-than-human geographies of today. 'Peak horse' also brought peak horse feed and peak horse dung, with all that meant for agricultural land use and cropping patterns and the stench of the urban environment. (It has been estimated that at its peak about a third of agricultural land was taken up by horses and growing horse feed.[53]) Peak horse also roughly coincided with the point of peak horse slaughter. Almost half a million horses were killed in the Boer War,[54] and the First World War brought untold equestrian carnage. The withdrawal of the use of horses in military fighting may have eased some of the worst excesses of the human–horse relationship in the twentieth century, but over that period interest in the welfare and material conditions of horses has changed significantly. The moral, political and ideological force behind the animal welfare movement is likely to be an increasingly important influence upon what happens to the human–horse relationship in future.

History canters towards better understanding the equestrian past while geography plods behind. Similarly, generating new insights into horse history is not being matched by attention to the future. Where are we going with horses, and who is envisioning equestrian futures? A study commissioned in the early 1980s by the Shire Horse Society is a remarkable piece of equine futurology that is evocative of another age. It considered the contemporary uses of heavy working horses and sought to promote their greater use. A Foreword by the Duke of Edinburgh set out the rationale for objectively thinking about how horses could be used more.

> There has been considerable advance in the design of machinery since the days when horses were last used to pull vehicles and farm implements.

> Horses cannot be made to go any faster, but modern technology has made it possible for them to achieve a substantial improvement in their efficiency as a source of motive power. If this report does nothing else, I hope it will make you think.[55]

Yet since the 1980s, heavy horses have not grown in use. They may continue to be a popular feature of rural heritage shows and museums, but their numbers remain low. They are precariously 'rare breeds'. Presciently, and despondently, the review noted that "the horse industry is too fragmented to demonstrate its own importance".[56] It recommended a federation of heavy horse societies in the hope that this might better serve the interests of those committed to a positive future for heavy working horses but warned of the need to avoid "cranks and welfarists" getting too involved.[57] Four decades on, what forces at play will mould and shape the future of horses in Britain's more-than-human geography?

In 2013, French researchers published an analysis of the future prospects of the horse in France. Through a foresight exercise, it considered 40 drivers of change likely to influence the future of the French horse sector, grouped into five themes: the economic and societal context; public expectations of horses and of equestrian and horseracing activities; public policy and regulation; equestrianism and horseracing; and the organisation and strategies of horse producers.[58] The study developed contrasting scenarios for the French horse industry in 2030. The first, called 'Everyone on Horseback', saw growing participation in equestrian activities, supported by economic growth and increased middle-class spending power. The second, called 'the High Society Horse', saw a more socially exclusive sector and declining horse numbers. The third, called 'the Civic Horse', saw a growing sector driven by local and regional public support, rooted in 'back-to-nature' interests in quality of life. The final scenario, called 'the Companion Horse', also saw a decline in horse numbers and a retreat from horse racing with the emphasis on horses as individual companion animals. Such detailed foresight work has not been carried out in the British context, but the drivers are broadly similar. The changing context for the sector over the past decade suggests that the British equestrian world is more likely to be on a path to a shrinking and more socially exclusive horse world than any of the other scenarios. Three driving forces appear most significant.

The first is economic. Difficult economic times for the country mean difficult times for the horse economy. The British Horseracing Authority reported that racehorse ownership fell by 10 per cent in the aftermath of the 2008 financial crisis. Total horse numbers in the UK are estimated to have fallen by over two-fifths since 2006, and the number of horse-owning households has fallen by a quarter since 2015.[59] In a country where disposable incomes are being squeezed, significant growth in the horse economy looks less propitious in 2023 than it has at any time in this century. The experience of Covid lockdowns may have spawned an increasing appetite for 'staycations' in the UK and greater seasonal demand for rural tourism and visitor attractions, but it is doubtful whether this will translate into rising horse numbers. The horseracing industry talks in terms of perpetual economic challenges and is dependent on money being paid by rich owners and via a hypothecated levy to keep it ticking over. The long-term structural shift towards online betting and betting on football suggests that horseracing will continue to be squeezed. It is an ageing sport that frets about how to bring in young (human) blood. Associated with economic pressures are institutional complexity and the different social and institutional bases for different parts of the horse world, especially racing and recreational horses. Despite efforts to improve co-ordination through the British Horse Council, the social and financial fissures between racing and the rest make the institutional representation of 'the horse sector' or 'the horse industry' an awkward construct, restricted to lobbying only on those limited common issues such as horse identification, trade and transport rules and disease threats. In addition are the social values of horse society and the ways these may change and differentiate over time. The tensions between social inclusion and exclusivity, or between cosmopolitanism and social conservatism, run through British society and politics, and the horse sector is no exception. These tensions in values come to a head around issues such as efforts to widen participation in equestrian activities. The sociological question of whether the people involved with horses will become more or less socially diverse is rooted in the economic question of the national economy and household finances. In an economic squeeze, it is likely that participation in horse sport and leisure activities will become more rather than less socially exclusive.

Considering the evolution of the horse world over a period of decades and centuries, among the strongest forces likely to affect the future of the horse in Britain is the increasing potency of animal welfare concerns. What in the 1980s might have been dismissed as the "cranks and welfarists" are far larger in number and increasingly armed with a growing body of scientific research on the welfare of the horse and the practices and environments that might compromise that welfare. Horseracing and other elite equestrian sports are under increasing pressure from an animal welfare perspective. It was the animal welfare movement that eventually forced the end of hunting wild mammals with hounds, though more out of a concern for the fox than for the horse. Similarly, whipping and horse casualties in jump racing, and the stresses placed on horses in Olympic equestrian sports, are having a bright light shone on them by committed campaigners. YouTube footage and social media outrage are bringing pressure to bear and forcing reviews and rethinks by governing bodies on the defensive. Thinking about the social license to operate means taking seriously the concerns people raise. A heightened animal welfare ethic is also reshaping the horse economy and beginning to alter the management of the lifecycle of the horse. There has been a growth in the number of equine osteopaths, masseurs and chiropractors and specialists who can advise on horse behaviour. Horses are increasingly seen as companion animals, and advances in veterinary science means they can live long beyond their 'useful years'. The concept of horse retirement may have been an anathema to nineteenth-century sensibilities, but an emergent form of livery are retirement yards that look after horses during the final part of the lives. Older horses are looked after by yard managers and staff who monitor and manage the final period of a horse's life and help ensure a good death.[60]

Alongside animal welfare, the rise of environmental concern is likely to be an important influence on how the relationship between humans and horses evolves in future. The implications of sustainability for horses and human–horse relationships are only beginning to be considered, both by equestrian organisations and by academic researchers.[61] Owning and living close to horses are often intertwined with notions of a post-materialist 'good life' in the country, and some have argued that domestic animals such as horses are members of a broader moral community and need to be incorporated into thinking about the root causes of unsustainable practices and modes of living.[62] The climate change challenge, however, is bringing a much more instrumentalist view of emissions reduction, carbon budgeting and transitioning to a net zero economy by 2050. This has particularly acute implications for rural land use and land management practices and is likely to lead to increasing pressures upon rural land for growing food, growing energy crops and delivering stretching targets for tree planting and carbon sequestration.[63] The current consideration of the horse world's role in sustainability is highly superficial, focusing on the use of single-use plastics, recyclable materials and so on. It takes the need for almost a million horses and as much as half a million hectares of land to service them as a given. As the climate crisis and the net zero commitment start to bite deeper, food prices and food security, climate change mitigation and other uses of land may come into more acute tension and conflict with using land for equestrian pleasures. More fundamental questions may come to be asked about the climate change footprint of the horse, adding a new dimension to the equestrian entanglements of twenty-first-century Britain's more-than-human geography.

Notes

1 Latour, B. (1993) *We Have Never Been Modern*. Hemel Hempstead: Harvester Wheatsheaf.
2 Haraway, D. (2008) *When Species Meet*. Minneapolis: University of Minnesota Press, p.301.
3 Haraway (2008), p.287.
4 Hill, C. (1988) *Horse Power: The Politics of the Turf*. Manchester: Manchester University Press.
5 Derry, M. (2006) *Horses in Society: A Story of Animal Breeding and Marketing Culture 1800–1920*. Toronto: Toronto University Press.
6 Anthony, D. (2007) *The Horse, The Wheel and Language: How Bronze-Age Riders from the Eurasian Steppes Shaped the Modern World*. Princeton: Princeton University Press.

7 Kelekna, P. (2009) *The Horse in Human History*. New York: Cambridge University Press.
8 Mitchell, P. (2015) *Horse Nations: The Worldwide Impact of the Horse on Indigenous Societies Post-1492*. Oxford: Oxford University Press.
9 Birke, L. and Thompson, K. (2018) *(Un)Stable Relations: Horses, Humans and Social Agency*. London: Routledge.
10 Raulff, U. (2018) *Farewell to the Horse: The Final Century of Our Relationship*. London: Penguin, p.7.
11 Guest, K and Mattfeld, M. (eds.) (2019a) *Horse Breeds and Human Society: Purity, Identity and the Making of the Modern Horse*. London: Routledge.
12 Guest, K. and Mattfeld, M. (eds.) (2019b) *Equestrian Cultures: Horses, Human Society and the Discourse of Modernity*. Chicago: University of Chicago Press.
13 Landry, D. (2009) *Noble Brutes: How Eastern Horses Transformed English Culture*. Baltimore: Johns Hopkins Press, quoted in Guest and Mattfeld (2019b), p.2.
14 An exception involving a geographer collaborating with veterinary scientists in a study of horse welfare is: Horseman, S. *et al.* (2016) Current welfare problems facing horses in Great Britain as identified by equine stakeholders, *PLoS One* 11, e0160269.
15 Mitchell (2015).
16 Hovorka, A. (2017) Animal geographies I: Globalizing and decolonizing. *Progress in Human Geography* 41, 382–94.
17 Barua, M. (2021) Infrastructure and non-human life: a wider ontology, *Progress in Human Geography* 45, 1467–89, p.1467.
18 Guest and Mattfeld (2019b), p.4.
19 Raulff (2018), p.11.
20 Guest and Mattfeld (2019b), p.4.
21 Forrest, S. (2016) *The Age of the Horse: An Equine Journey through Human History*. London: Atlantic; Raulff (2018).
22 Greene, A. (2008) *Horses at Work: Harnessing Power in Industrial America*. Cambridge, MA: Harvard University Press.
23 McShane, C. and Tarr, J. (2007) *The Horse in the City: Living Machines in the Nineteenth Century*. Baltimore: Johns Hopkins Press.
24 Almeroth-Williams, T. (2019) *City of Beasts: How Animals Shaped Georgian London*. Manchester: Manchester University Press.
25 Swart, S. (2007) But where's the bloody horse?: Textuality and corporeality in the "animal turn", *Journal of Literary Studies*, 23, 271–292; Swart, S. (2010) *Riding High: Horses, Humans and History in South Africa*. Johannesburg: University of Witwatersrand.
26 Chandra, Y. (2022) *The Tale of the Horse: A History of India on Horseback*. Newbury: Holland House.
27 Raulff (2018), p.8.
28 House of Commons Environment, Food and Rural Affairs Committee (2013) *Contamination of Beef Products*. Session 2012–13. London: Stationary Office.
29 Fairlie, S. (2010) *Meat: A Benign Extravagance*. East Meon: Permanent Publications; Oliver, C. (2021) *Vegansim, Archives and Animals: Beyond Human Geographies*. London: Routledge; Percival, R. (2022) *The Meat Paradox: Eating, Empathy and the Future of Meat*. London: Little Brown.
30 See the following for reflections on the issue: Otter, C. (2011) Hippophagy in the UK: a failed dietary revolution, *Endeavour* 35, 80–90; Forrest, S. (2019) 'Horsemeat is certainly delicious': Anxiety, zenophobia, and rationalism at a nineteenth century American hippophagic banquet, pp.160–78 in K. Guest, K. and M. Mattfeld (eds.) (2019) *Equestrian Cultures: Horses, Human Society and the Discourse of Modernity*. Chicago: University of Chicago Press.
31 Otter (2011), p.80.

32 Otter (2011), p.82.
33 Otter (2011), p.83.
34 Otter (2011), p.86.
35 Otter (2011), p.86.
36 House of Commons Environment, Food and Rural Affairs Committee (2013), p.7.
37 Taylor, J. (2022) *'I Can't Watch Anymore': The Case for Dropping Equestrian from the Olympic Games*. Copenhagen: Epona Publishing, p.5.
38 Taylor (2022), p.7.
39 Taylor (2022), p.23. The 70 per cent figure does not include those riders who are registered with federations outside Europe but who have to live in Europe in order to compete.
40 Mitchell (2015).
41 Barker, T. (1988) Urban transport, in M. Freeman and D. Aldcroft (eds.) *Transport in Victorian Britain*, Manchester: Manchester University Press, p.134, quoted in M. Doole (2020) Identification of the urban infrastructure of nineteenth-century horse transport: a case study of Worksop, Nottinghamshire, UK, pp.191–206 in D. Turner (ed.) *Transport and Its Place in History: Making the Connections*. London: Routledge.
42 Thompson, F.M.L. (1976) Nineteenth century horse sense, *Economic History Review* 29, 60–81.
43 Ensminger, E. (1969). *Horses and Horsemanship*. Danville, Ill.: Interstate Printers and Publishers.
44 Swart, S. (2010) *Riding High: Horse, Humans and History in South Africa*. Johannesburg: University of Witwatersrand, p.143.
45 Edgerton, D. (2019) *The Shock of the Old: Technology and Global History since 1900*. (Second edition) London: Profile, p.33.
46 Doole, M. (2020) p.195.
47 Paxman, J. (2021) *Black Gold: The History of How Coal Made Britain*. London: William Collins, p.18.
48 Edgerton (2019).
49 Chivers, K. (1988) *History with a Future: Harnessing the Heavy Horse for the Twenty-First Century*. Peterborough: The Shire Horse Society, p.12.
50 Examples include: Ebony Horse Club in Brixton, Park Palace Ponies in Liverpool, Stepney Bank in Newcastle (Newcastle Inner City Community Horse Project), Wormwood Scrubbs Pony Club, Vauxhall City Farm, Shy Lowen Horse and Pony Sanctuary in Liverpool, Greatwood Horse Power Programme (Wiltshire), Grassroots Urban Horse and Pony Club (Knowle West Estate, Bristol), Emile Faurie Foundation, Docklands Pony Club, Park Lane Stables in Teddington, West London Community Riding Centre.
51 Edwards, P. and Graham, E. (2011) Introduction, pp.1–33 in P. Edwards *et al.* (eds.) *The Horse as Cultural Icon: The Real and Symbolic Horse in the Early Modern Period*. Leiden: Brill, p.1.
52 Lorimer, J. (2020) *The Probiotic Planet: Using Life to Manage Life*. Minneapolis: University of Minnesota Press; see also Hinchliffe, S. *et al.* (2016) *Pathological Lives: Disease, Space and Biopolitics*. Chichester: Wiley.
53 Edgerton (2019), pp.33–4.
54 Swart (2010), p.137.
55 Chivers (1988), pp.iii–vi.
56 Chivers (1988), p.10.
57 Chivers (1988), p.189.
58 Jez, C. *et al.* (2013) Factors driving change in the French horse industry to 2030, *Advances in Animal Biosciences* 4:s2, 66–105; Jez, C. *et al.* (2013) Scenarios, *Advances in Animal Biosciences* 4:s2, 106–15.

59 JDA Research (2023) *The National Equestrian Survey 2023. Overview Report.* Wakefield: British Equestrian Trade Association; Jones, E. (2023) Riding at risk as schools close but participation slightly up, *Horse & Hound*, 9 March p.4.
60 Schuurman, N. and Franklin, A. (2018) A good time to die: Horse retirement yards as shared spaces of interspecies care and accomplishment, *Journal of Rural Studies* 57, 110–117.
61 Wadham, H. (2020) Horse matters: re-examining sustainability through human-domestic animal relationships, *Sociologia Ruralis* 60, 530–50; Wadham, H. *et al.* (2023) Agents of sustainability: How horses and people co-create, enact and embed the good life in rural places, *Sociologia Ruralis* 63, 390–414.
62 Wadham (2020), p.546.
63 Ward, N. (2023) *Net Zero, Food and Farming: Climate Change and the UK Agri-Food System.* London: Routledge.

Bibliography

Allen, R. (2009) *The British Industrial Revolution in Global Perspective*. Cambridge: Cambridge University Press.

Almeroth-Williams, T. (2013) The brewery horse and the importance of equine power in Hanoverian London, *Urban History* 40, 416–441. doi:10.1017/S0963926813000333

Almeroth-Williams, T. (2019) *City of Beasts: How Animals Shaped Georgian London*. Manchester: Manchester University Press.

Anderson, V. (2004) *Creatures of Empire: How Domestic Animals Transformed Early America*. Oxford: Oxford University Press.

Andersson, P. (2019) Who is the horse? Horse assisted psychotherapy as a possibility for understanding horses, pp.45–56 in J. Bornemark, P. Andersson and U. von Essen (eds.) *Equine Cultures in Transition: Ethical Questions*. London: Routledge.

Animal Aid (2008) *Race Horse Death Watch: The First Year*. Tonbridge: Animal Aid. https://www.horsedeathwatch.com/downloads/DeathWatchone.pdf (Accessed 1 July 2023)

Animal Aid (2020) *186 Race Horses Killed in 2019*. Tonbridge: Animal Aid https://www.horsedeathwatch.com/ (Accessed 1 July 2023)

Anthony, D. (2007) *The Horse, The Wheel and Language: How Bronze-Age Riders from the Eurasian Steppes Shaped the Modern World*. Princeton: Princeton University Press.

Austen, B. (1978) *English Provincial Posts, 1633-1840*. London: Phillimore.

Bachrach, B. (1985) On the origins of William the Conqueror's horse transports, *Technology and Culture* 26, 505–31. doi:10.2307/3104851

Baird, W. and Tarrant, J. (1973) *Hedgerow Destruction in Norfolk, 1946-1970*. Centre of East Anglian Studies. Norwich: University of East Anglia.

Bankes, C. (2014) Half of all sport horse are lame, study finds, *Horse & Hound* 28 April 2014 https://www.horseandhound.co.uk/news/half-horses-lame-saddle-slip-survey-428728 (Accessed 1 July 2023)

Barker, R. (1983) The delayed decline of the horse in the twentieth century, pp.101–112 in F.M.L. Thompson (ed.) *Horses in European Economic History: A Preliminary Canter*. Reading: British Agricultural History Society.

Barker, T. (1988) Urban transport, in M. Freeman and D. Aldcroft (eds.) *Transport in Victorian Britain*, Manchester: Manchester University Press

Barker, T. and Gerhold, D. (1993) *The Rise and Rise of Road Transport, 1700–1900*. Basingstoke: Macmillan.

Barua, M. (2021) Infrastructure and non-human life: a wider ontology, *Progress in Human Geography* 45, 1467–89. doi:10.1177/0309132521991220

Bedford, M. (2017) Withen: The battle of Orgreave, 1984, pp.335–65 in R. Page (ed.) *Protest: Stories of Resistance*. Manchester: Comma Press.

Bell, C., Rogers, S., Taylor, J. and Busby, D. (2019) Improving the recognition of equine affective states, *Animals* 9, 1124, doi:10.3390/ani9121124

Bell, S. (2020) Horse racing in imperial Rome: Athletic competition, equine performance, and urban spectacle, *International Journal of the History of Sport* 37, 183–232. doi:10.1080/09523367.2020.1782385

Bevan, J. (2011) *Foxhunting and the Landscape between 1700 and 1900 With Particular Reference to Norfolk and Shropshire*. PhD Thesis, University of East Anglia, School of History. https://core.ac.uk/download/pdf/8780333.pdf (Accessed 1 July 2023).

Birke, L. and Thompson, K. (2018) *(Un)Stable Relations: Horses, Humans and Social Agency*. London: Routledge.

Blackburn, D. (2021) Equine-Assisted Therapy as complementary practice in the treatment of operational stress injuries of Canadian Armed forces veterans, *SunText Review of Neuroscience & Psychology* 2(1), 125 doi:10.51737/2766-4503.2021.025

Blomberg, J. and Välimaa, C. (2016) *Horses and Sustainable Development: How Can the Horse Become the Caretaker of the Planet?* Stockholm: Swedish Horse Industry Foundation.http://xn--hllbarhst-12ae.se/wp-content/uploads/2017/02/Report-sustainable-horse-2016.pdf (Accessed 1 July 2023).

Boden, L., Parkin, T., Yates, J., Mellor, D. and Kao, R. (2012) Summary of current knowledge of the size and spatial distribution of the horse population within Great Britain, *BMC Veterinary Research* 8:43. doi:10.1186/1746-6148-8-43

Bomans, K., Dew Aelheyns, V. and Gulinik, H. (2011) Pasture for horses: An underestimated land use class in an urbanised and multifunctional area, *International Journal of Sustainable Development and Planning* 6, 195–211. doi:10.2495/SDP-V6-N2-195-211

Brayshay, M. (1991) Royal post-horse routes in England and Wales: the evolution of the network in the later-sixteenth and early-seventeenth century, *Journal of Historical Geography* 17, 373–89. doi:10.1016/0305-7488(91)90023-O

Brayshay, M. (2014) *Land Travel and Communications in Tudor and Stuart England: Achieving a Joined-Up Realm*. Liverpool: Liverpool University Press.

Brayshay, M., Harrison, P. and Chalkley, B. (1998) Knowledge, nationhood and governance: the speed of the Royal post in early-modern England, *Journal of Historical Geography* 24, 265–88. doi:10.1006/JHGE.1998.0087

British Equestrian (2022) *Barriers to Equestrianism: Invitation to Tender*. Kenilworth: British Equestrian https://bef-admin.ideasbarn.com/assets/British_Equestrian_Research%20Brief_EDI_project.pdf (Accessed 1 July 2023)

British Equestrian Federation (2020a) Costs of travelling horses to Europe post-Brexit, *News Release* https://www.britishequestrian.org.uk/assets/NEWS/Cost%20comparision%20of%20travelling%20a%20horse%20to%20Europe%20FINAL.pdf (Accessed 1 July 2023)

British Equestrian Federation (2020b) *Enriching the Lives of People and Their Horses Together. Strategy 2020–2024*. Kenilworth: British Equestrian. https://www.britishequestrian.org.uk/assets/About%20the%20BEF/BEF_Strategy_Document_FINAL.pdf (Accessed 6 June 2022).

British Grooms Association and Equestrian Employers Association (2020) Coronavirus Impact Survey. https://britishgrooms.org.uk/news/451/the-impact-results (Accessed 1 July 2023).

British Horse Council (2019) *The British Horse Sector – Why it Matters for the 2019 General Election*. High Wycombe: British Horse Council.

British Horse Industry Confederation (2008) Memorandum of evidence, pp.169–71 in House of Common Environment, Food and Rural Affairs Committee (2008) *The*

Potential of England's Rural Economy. HC544-II. Session 2007/08. London: The Stationary Office https://publications.parliament.uk/pa/cm200708/cmselect/cmenvfru/544/544we15.htm (Accessed 1 July 2023)

British Horse Industry Confederation, Defra, Department for Culture, Media and Sport and the Welsh Assembly Government (2005) *Strategy for the Horse Industry in England and Wales*. London: Defra.

British Horseracing Authority (2018) *The Cheltenham Festival Review 2018*. London: British Horseracing Authority. https://www.britishhorseracing.com/wp-content/uploads/2018/12/Cheltenham-Festival-Review-2018.pdf (Accessed 1 July 2023).

British Horseracing Authority (2020a) Written Evidence to the House of Commons Culture, Media and Sport Committee Inquiry into *The Impact of Covid-19 on DCMS Sectors*. Session 2019–21. London: Stationary Office https://committees.parliament.uk/writtenevidence/7020/pdf/ (Accessed 1 July 2023)

British Horseracing Authority (2020b) *Annual Report and Consolidated Financial Statements*. London: BHA. http://media.britishhorseracing.com/bha/Publications/Annual_Reports/2020.pdf (Accessed 1 July 2023)

British Horseracing Authority (2021) Environmental sustainability project to review racing's progress and support long-term planning, *British Horseracing Authority Press Release* 7 December, https://www.britishhorseracing.com/press_releases/racings-sustainable-future-help-map-our-environmental-activity/ (Accessed 1 July 2023).

British Horseracing Authority (2022) Racing's sustainable future – help map our environmental activity, *British Horseracing Authority Press Release* 8 February, https://www.britishhorseracing.com/press_releases/racings-sustainable-future-help-map-our-environmental-activity/ (Accessed 1 July 2023).

Brown, J. (2008) *Steam on the Farm: A History of Agricultural Steam Engines 1800 to 1950*. Marlborough: Crowood Press.

Bucks, S. and Wadey, P. (2017) *Rights of Way: Restoring the Record* (Second edition) Ilminster: Bucks & Wadey Publishing.

Budiansky, S. (1997) *The Nature of Horses: Their Evolution, Intelligence and Behaviour*. London: Weidenfeld & Nicholson.

Buller, H. (2013) Animal geographies I, *Progress in Human Geography* 38, 308–18. doi:10.1177/0309132513479295

Buller, H. (2014) Animal geographies II: methods, *Progress in Human Geography* 39, 374–84. doi:10.1177/0309132514527401

Buller, H. (2015) Animal geographies III: ethics, *Progress in Human Geography* 40, 422–30. doi:10.1177/0309132515580489

Burgon, H. (2011) *A Reflexive Exploration of the Experiences of At-Risk Young People Participating in Therapeutic Horsemanship*. PhD Thesis, Cardiff University School of Social Sciences. https://orca.cardiff.ac.uk/54446/1/U584547.pdf (Accessed 1 July 2023).

Burgon, H., Gammage, D. and Hebden, J. (2018) Hoofbeats and heartbeats: equine-assisted therapy and learning with young people with psychological issues – theory and practice, *Journal of Social Work Practice* 32, 3–16 doi:10.1080/02650533.2017.1300878

Burns, D. (1992) *Poll Tax Rebellion*. Stirling: AK Press.

Campaign for the Protection of Hunted Animals (1998) Anglers and riders support Foster Bill to ban hunting with dogs, *Campaign for the Protection of Hunted Animals Press Release* 3 February, London: CPHA.

Campbell, J. (1967) *Police Horses*. Newton Abbott: David & Charles.

Campbell, M. (2019) *Animals, Ethics and Us: A Veterinary's View of Human-Animal Interactions*. Sheffield: 5m Books.

Campbell, M. and McNamee, M. (2021) Ethics, genetic technologies and equine sports: The prospect of regulation of a modified therapeutic use exemption policy, *Sport, Ethics and Philosophy* 15, 227–50. doi:10.1080/17511321.2020.1737204

Cassidy, R. (2002) *The Sport of Kings: Kinship, Class and Thoroughbred Breeding in Newmarket*. Cambridge: Cambridge University Press.

Cassidy, R. (2007) *Horse People: Thoroughbred Culture in Lexington and Newmarket*. Baltimore: Johns Hopkins University Press.

Cassidy, R. (2020) *Vicious Games: Capitalism and Gambling*. London: Pluto.

Chamberlain, J.E. (2007) *Horse: How the Horse Has Changed Civilisations*. Oxford: Signal.

Chandra, Y. (2022) *The Tale of the Horse: A History of India on Horseback*. Newbury: Holland House.

Chartres, J. (1985) The marketing of agricultural produce, pp.406–502 J. Thirsk (ed.) *The Agrarian History of England and Wales Volume V-II 1640–1750*. Cambridge: Cambridge University Press, pp.436–7.

Chartres, J. (2002) The eighteenth century English inn: a transient 'golden age'? pp.205–226 in B. Kumin and B. Tlusty (ed.) *The World of the Tavern: Public Houses in Early Modern Europe*. Basingstoke: Ashgate.

Chivers, K. (1988) *History with a Future: Harnessing the Heavy Horse for the Twenty-First Century*. Peterborough: The Shire Horse Society.

Clayton, M., Pritchard-Gordon, G. and Dodd, A. (1999) Oral evidence, House of Commons Select Committee on Environment, Transport and Regional Affairs *The Rural White Paper*, Session 1999–2000, HC Paper 32, London: The Stationery Office https://publications.parliament.uk/pa/cm199900/cmselect/cmenvtra/32/9121508.htm (Accessed 1 July 2023).

Climate Change Committee (2020) *Land Use: Policies for a Net Zero UK*. London: Climate Change Committee. https://www.theccc.org.uk/publication/land-use-policies-for-a-net-zero-uk/ (Accessed 1 July 2023)

Cobham Resource Consultants (1997) *Countryside Sports - Their Economic, Social and Conservation Significance*, Reading: Standing Conference on Countryside Sports.

Cockayne, E. (2007) *Hubbub: Filth, Noise and Stench in England 1600–1770*. London: Yale University Press.

Colley, L. (1992) *Britons: Forging the Nation 1707–1837*. London: Pimlico.

Collins, E. (1983) The farm horse economy of England and Wales in the early tractor age 1900–40, pp.73–100 in F.M.L. Thompson (ed.) *Horses in European Economic History: A Preliminary Canter*. Reading: British Agricultural History Society.

Collins, E. (2000) *The Agrarian History of England and Wales Volume VII 1850–1914*, Parts I & II. Cambridge University Press.

Committee of Inquiry into Hunting with Dogs in England and Wales (2000) *Report of the Committee of Inquiry into Hunting with Dogs in England and Wales* Cm4763. London: Home Office.

Conn, D. (2017) The scandal of Orgreave, *The Guardian* Long Read, 18 May 2017. https://www.theguardian.com/politics/2017/may/18/scandal-of-orgreave-miners-strike-hillsborough-theresa-may (Accessed 1 July 2023).

Creswell, H. (1958) Seventy years back, *Architectural Review*, December 1958

Crossman, G. (2010) *The Organisational Landscape of the English Horse Industry: A Contrast with Sweden and the Netherlands*. University of Exeter PhD thesis. https://ethos.bl.uk/OrderDetails.do?uin=uk.bl.ethos.546971 (Accessed 1 July 2023).

Crossman, G. and Walsh, R. (2011) The changing role of the horse: From beast of burden to partner in sport and recreation, *International Journal of Sport and Society* 2, 95–110. doi:10.18848/2152-7857/CGP/v02i02/54066

Dashper, K., Abbott, J. and Wallace, C. (2020) 'Do horses cause divorces?' Autoethnographic insights on family, relationships and resource-intensive leisure, *Annals of Leisure Research* 23, 304–21. doi:10.1080/11745398.2019.1616573

Davies, E., McConn-Palfreyman, W., Williams, J. and Lovell, G. (2020) The impact of Covid-19 on staff working practices in UK horseracing, *Animals* 10(11) 2003. doi:10.3390/ani10112003

de Belin, M. (2013) *From the Deer to the Fox: The Hunting Tradition and the Landscape 1600–1850*. Hatfield: University of Hertfordshire Press.

de Vries, J. (1984) *European Urbanisation 1500–1800*. London: Routledge.

Deloitte (2013) *Economic Impact of British Horseracing 2013*. London: British Horseracing Authority. https://www.britishhorseracing.com/wp-content/uploads/2014/03/EconomicImpactStudy2013.pdf (Accessed 1 July 2023).

Demuijnck, G. and Fasterling, B. (2016) The social licence to operate, *Journal of Business Ethics* 136, 675–85 doi:10.1007/s10551-015-2976-7

Department of Sustainability, Environment, Water, Population and Communities [Australian Government] (2011) *Feral Horse (Equus Caballus) and Feral Donkey (Equus Asinus)*. Canberra: Department of Sustainability, Environment, Water, Population and Communities. https://www.agriculture.gov.au/sites/default/files/documents/feral-horse.pdf (Accessed 1 July 2023).

Derry, M. (2006) *Horses in Society: A Story of Animal Breeding and Marketing Culture 1800–1920*. Toronto: Toronto University Press.

Donoughue, B. (2016) *Westminster Diary Volume 1: A Reluctant Minister Under Tony Blair*. London: I.B. Tauris.

Donoughue, B. (2018) *Westminster Diary Volume 2: Farewell to Office*. London: I.B. Tauris.

Doole, M. (2020) Identification of the urban infrastructure of nineteenth-century horse transport: a case study of Worksop, Nottinghamshire, UK, pp.191–206 in D. Turner (ed.) *Transport and Its Place in History: Making the Connections*. London: Routledge.

Douglas, J., Owers, R. and Campbell, M. (2022) Social licence to operate: What can equestrian sports learn from other industries, *Animals* 12, 1987. doi:10.3390/ani12151987

Drury, J. (2017) The poll tax riot – afterword: from peaceful protest to riot, 31 March 1990, pp.422–27 in R. Page (ed.) *Protest: Stories of Resistance*. Manchester: Comma Press.

Duncan, E., Graham, R. and McManus, P. (2018) 'No one has ever seen … smelt … or sensed a social licence': Animal geographies and social licence to operate, *Geoforum* 96, 318–327. doi:10.1016/j.geoforum.2018.08.020

East, R., Power, H. and Thomas, P. (1985) The death of mass picketing, *Journal of Law and Society* 12, 305–19 doi:10.2307/141024

Edgerton, D. (2019) *The Shock of the Old: Technology and Global History since 1900*. (Second edition) London: Profile.

Edwards, P. (1988) *The Horse Trade of Tudor and Stuart England*. Cambridge: Cambridge University Press.

Edwards, P. (2007) *Horse and Man in Early Modern England*. London: Hambledon Continuum.

Edwards, P. and Graham, E. (2011) Introduction, pp.1–33 in P. Edwards, K. Enenkel and E. Graham (eds.) *The Horse As Cultural Icon: The Real and Symbolic Horse in the Early Modern Period*. Leiden: Brill.

Elder, L. (2022) How the economy is affecting the horse world, *Horse & Hound* 29 September, pp.4–5 https://www.horseandhound.co.uk/publication/horse-and-hound-magazine/horse-hound-29-september-2022 (Accessed 1 July 2023).

Elghandour, E., Adegbeye, M., Barbabose-Pilego, A., Perez, N., Hernández, S., Zaragoza-Bastida, A. and Salem, A. (2019) Equine contribution in methane emission and its mitigation strategies, *Journal of Equine Veterinary Science* 72, 56–63 doi:10.1016/j.jevs.2018.10.020

Ellis, J. (2004) *Cavalry: The History of Mounted Warfare*. Barnsley: Pen and Sword Books.

Emel, J., Wilbert, C. and Wolch, J. (2002) Animal geographies, *Society & Animals* 10, 407–12. doi:10.1163/156853002320936881

Ensminger, E. (1969). *Horses and Horsemanship*. Danville, Ill: Interstate Printers and Publishers

Equestrian Employers' Association (2022a) *National Minimum Wage Rise Impact Survey*. Tarporley: Equestrian Employers' Association https://britishgrooms.org.uk/uploads/files/EEA/EEA_NMW%20impact%20survey%20results%20report.pdf (Accessed 1 July 2023).

Equestrian Employers' Association (2022b) 71% concerned about viability, *News Release* 3rd March 2022 https://equestrianemployers.org.uk/news/546/71-concerned-about-viability (Accessed 1 July 2023).

Equine Ethics and Wellbeing Commission (2022) *Public Attitudes on the Use of Horses in Sport: Survey Report*. https://equinewellbeing.fei.org/assets/documents/Results%20of%20General%20Public%20Survey%20-%20%20Equine%20Ethics%20and%20Wellbeing%20Commission%20Report%202022.pdf (Accessed 1 July 2023).

European Horse Network (2017) *Equestrian Tourism*. Briefing Note, July 2017. Brussels: European Horse Network https://www.europeanhorsenetwork.eu/documents/position-statements/ (Accessed 1 July 2023).

Everitt, A. (1967) Farm labourers, pp.396–465 in J. Thirsk (ed.) *The Agrarian History of England and Wales Volume IV 1500-1640*. Cambridge: Cambridge University Press.

Fairlie, S. (2010) *Meat: A Benign Extravagance*. East Meon: Permanent Publications.

Ford, R. and Goodwin, M. (2014) *Revolt on the Right: Explaining Support for the Radical Right in Britain*. London: Routledge.

Forrest, S. (2016) *The Age of the Horse: An Equine Journey through Human History*. London: Atlantic.

Forrest, S. (2019) 'Horsemeat is certainly delicious': Anxiety, zenophobia, and rationalism at a nineteenth century American hippophagic banquet, pp.160–78 in K. Guest, K. and M. Mattfeld (eds.) (2019) *Equestrian Cultures: Horses, Human Society and the Discourse of Modernity*. Chicago: University of Chicago Press

Fox, K. (1999) *The Racing Tribe: Watching the Horsewatchers*. London: Metro Books.

Francis-Baker, T. (2023) *The Bridleway: How Horses Shaped the British Landscape*. London: Bloomsbury.

Franzen, J. (2010) *The Rise of the Horse: 55 Million Years of Evolution*. Baltimore: Johns Hopkins University Press.

Fraser, T. (2019) *Livestock and Landscape: Livestock Improvement and Landscape Enclosure in Late and Post-Medieval England*, PhD Thesis, University of Sheffield, Department of Archaeology. https://etheses.whiterose.ac.uk/27470/1/T_Fraser_Livestock%20_and_Landscape.pdf (Accessed 1 July 2023).

Frontier Economics (2016) *An Economic Analysis of the Funding of Horseracing*. Report to the Department for Culture, Media and Sport. London: Frontier Economics https://assets.publishing.service.gov.uk/government/uploads/system/uploads/attachment_data/file/586305/Frontier_Economics-_An_economic_analysis_of_the_funding_of_horseracing.pdf (Accessed 1 July 2023).

Furtado, T. (2019) *Exploring the Recognition and Management of Obesity in Horses through Qualitative Research*. PhD Thesis. University of Liverpool. https://core.ac.uk/download/pdf/228137274.pdf

Furtado, T., King, M., Perkins, E., McGowan, C., Chubbock, S., Hannelly, E., Rogers, J. and Pinchbeck, G. (2022) An exploration of environmentally sustainable practices associated with alternative grazing management system use for horses, ponies, donkeys and mules in the UK, *Animals* 12, 151. doi:10.3390/ani12020151

Furtado, T., Perkins, E., McGowan, C. and Pinchbeck, G. (2021c) Equine management in UK livery yards during the COVID-19 pandemic — "As long as the horses are happy, we can work out the rest later", *Animals* 11, 1416. doi:10.3390/ani11051416

Furtado, T., Perkins, E., McGowan, C., Watkins, F., Pinchbeck, G. and Christly, R. (2018) *When the Grass is Greener: The Equine Weight Management Guide for Every Horse, Every Yard, and Every Owner*. Princes Risborough: The Horse Trust and University of Liverpool.

Furtado, T., Perkins, E., Pinchbeck, G., Watkins, F. and Christley, R. (2021a) Exploring horse owners' understanding of obese body condition and weight management in UK leisure horses, *Equine Veterinary Journal* 53, 752–62. doi:10.1111/evj.13360

Furtado, T., Perkins, E., Pinchbeck, G., Watkins, F. and Christley, R. (2021b) Hidden in plain sight: Uncovering the obesogenic environment surrounding the UK's leisure horses, *Antrozoös* 34, 491–506, doi:10.1080/08927936.2021.1914431

Garner, R. (1993) *Animals, Politics and Morality*. Manchester: Manchester University Press.

Garner, R. (2002) Political science and animal studies, *Society and Animals* 10, 395–401. doi:10.1163/156853002320936863

Geels, F. (2005) The dynamics of transitions in socio-technical systems: A multi-level analysis of the transition pathway from horse-drawn carriages to automobiles (1860–1930), *Technology Analysis & Strategic Management* 17, 445–76. doi:10.1080/09537320500357319

Gehman, J., Lefsrud, L. and Fast, S. (2017) Social licence to operate: legitimacy by another name, *Canadian Public Administration* 60, 293–317. doi:10.1111/capa.12218

George, J. (1999) *A Rural Uprising: The Battle to Save Hunting with Hounds* London: J. A. Allen.

Giacomantonio, C., Bradford, B., Davies, M. and Martin, R. (2015) *Making and Breaking Barriers: Assessing the Value of Mounted Police Units in the UK.* Cambridge: RAND Europe. https://www.rand.org/pubs/research_reports/RR830.html (Accessed 1 July 2023).

Gibbs, L (2020) Animal geographies I: Hearing the cry and extending beyond. *Progress in Human Geography* 44, 769–77. doi:10.1177/0309132519863483

Gibbs, L (2021) Animal geographies II: Killing and caring in times of crisis. *Progress in Human Geography* 45, 371–81. doi:10.1177/0309132520942295

Glennie, P. and Thrift, N. (2009) *Shaping the Day: A History of Timekeeping in England and Eales, 1300–1800*. Oxford: Oxford University Press.

Greene, A. (2008) *Horses at Work: Harnessing Power in Industrial America*. Cambridge, MA: Harvard University Press.

Griffin, E. (2007) *Blood Sport: Hunting in Britain Since 1066*. Yale University Press.

Guest, K. and Mattfeld, M. (eds.) (2019a) *Horse Breeds and Human Society: Purity, Identity and the Making of the Modern Horse*. London: Routledge.

Guest, K. and Mattfeld, M. (eds.) (2019b) *Equestrian Cultures: Horses, Human Society and the Discourse of Modernity*. Chicago: University of Chicago Press.

Hämäläinen, P. (2003) The rise and fall of plains Indian horse culture, *Journal of American History* 90, 833–62. doi:10.2307/3660878
Haraway, D. (2008) *When Species Meet*. Minneapolis: University of Minnesota Press.
Hartpury University (2020) Research reinforces benefits of equine-assisted therapy for children, *Press Release* 17 August 2020, https://www.hartpury.ac.uk/news/2020/08/research-reinforces-benefits-of-equine-assisted-therapy-for-children/ (Accessed 1 July 2023).
Harvey, D. (1996) *Justice, Nature and the Geography of Difference*. Oxford: Blackwells.
Harvey, E. (2010) Pavage grants and urban street paving in medieval England, 1249–1462, *Journal of Transport History* 31, 151–63. doi:10.7227/TJTH.31.2.3
Harvey, F. (2013) Huge rise in first time hunters for new foxhunting season, *The Guardian* 2nd November, p. 18.
Havukainen, J., Väisänen, S., Rantala, T., Saunila, M. and Ukko, J. (2020) Environmental impacts of manure management based on life cycle assessment approach, *Journal of Cleaner Production* 264, 121576. doi:10.1016/j.jclepro.2020.121576 0959-6526
Hill, C. (1988) *Horse Power: The Politics of the Turf*. Manchester: Manchester University Press.
Hinchliffe, S., Bingham, N., Allen, J. and Carter, S. (2016) *Pathological Lives: Disease, Space and Biopolitics*. Chichester: Wiley.
Hobsbawm, E. (1962) *The Age of Revolution 1789–1848*. London: Weidenfeld and Nicholson.
Hobsbawm, E. (1975) *The Age of Capital 1848–1875*. London: Weidenfeld and Nicholson.
Hobsbawm, E. (1987) *The Age of Empire 1875–1914*. London: Weidenfeld and Nicholson.
Holderness, B. (1989) Prices, productivity and output, pp.84–274 in G. Mingay (ed.) *The Agrarian History of England and Wales, Volume VI, 1750–1850*. Cambridge: Cambridge University Press.
Horse Welfare Board (2020) *A Life Well Lived: A New Strategic Plan for the Welfare of Horses Bred for Racing 2020–2024*. London: British Horseracing Authority. http://media.britishhorseracing.com/bha/Welfare/HWB/WELFARE_STRATEGY.pdf (Accessed 1 July 2023)
Horseman, S., Buller, H., Mullan, S. and Whay, H. (2016) Current welfare problems facing horses in Great Britain as identified by equine stakeholders, *PLoS One* 11, e0160269. doi:10.1371/journal. pone.0160269
Horserace Betting Levy Board (2023) *Annual Report and Accounts 2021/22*. London: HBLB. https://www.hblb.org.uk/documents/Executive/HBLB%20Annual%20Report%20and%20Accounts%202021-22.pdf (Accessed 1 July 2023).
Hoskins, W. (1955) *The Making of the English Landscape*. Harmondsworth: Penguin.
House of Commons Environment, Food and Rural Affairs Committee (2013) *Contamination of Beef Products*. Session 2012–13. London: Stationary Office. https://publications.parliament.uk/pa/cm201213/cmselect/cmenvfru/946/946.pdf (Accessed 1 July 2023).
House of Commons Environment, Food and Rural Affairs Committee (2021) *Moving Animals Across Borders*. HC Paper 79, Session 2021/22 London: Stationary Office. https://committees.parliament.uk/publications/7464/documents/78318/default/ (Accessed 1 July 2023).
Hovorka, A. (2017) Animal geographies I: Globalizing and decolonizing. *Progress in Human Geography* 41, 382–94. doi:10.1177/0309132516646291

Hovorka, A. (2018) Animal geographies II: Hybridizing. *Progress in Human Geography* 42, 453–62. doi:10.1177/0309132517699924

Hovorka, A. (2019) Animal geographies III: Species relations of power. *Progress in Human Geography* 43, 749–57. doi:10.1177/0309132518775837

Hovorka, A., McCubbin, S., and van Patter, L. (eds.) (2021a) *A Research Agenda for Animal Geographies*. Cheltenham: Edwin Elgar.

Hovorka, A., McCubbin, S., and van Patter, L. (2021b) Introduction to A Research Agenda for Animal Geographies: Visioning amidst socio-ecological crises, pp.1–20 in A. Hovorka et al. (eds.) (2021a) *A Research Agenda for Animal Geographies*. Cheltenham: Edwin Elgar.

Huey, R. (2022) What's the value of a horse?: An economic strategy for the equine sector in Northern Ireland, presentation to the National Equine Forum, 3rd March 2022 https://www.nationalequineforum.com/forum-2022/ (Accessed 22 March 2023).

Huggins, M. (2018) *Horse Racing and British Society in the Long Eighteenth Century*. Woodbridge: Boydell Press.

Hyland, A. (1990) *Equus: The Horse in the Roman World*. London: Batsford.

Itzkowitz, D. (2016) *Peculiar Privilege: A Social History of English Foxhunting, 1753-1885*. Brighton: Edward Everard Root.

James, J. (1994) *Debt of Honour: History of the International League for the Protection of Horses*. Basingstoke: Macmillan.

JDA Research (2019) *The National Equestrian Survey 2019. Overview Report*. Wakefield: British Equestrian Trade Association.

JDA Research (2023) *The National Equestrian Survey 2023. Overview Report*. Wakefield: British Equestrian Trade Association.

Jez, C., Coudurier, B., Cressent, M. and Méa, F. (2013a) Factors driving change in the French horse industry to 2030, *Advances in Animal Biosciences* 4:s2, 66–105. doi:10.1017/S2040470013000368

Jez, C., Coudurier, B., Cressent, M. and Méa, F. (2013b) Scenarios, *Advances in Animal Biosciences* 4:s2, 106–15. doi:10.1017/S204047001300037X

Jones, B., Goodfellow, J., Yeates, J. & McGreevy, P. (2015) A critical analysis of the British Horseracing Authority's review of the use of the whip in horseracing. *Animals* 5: 138–150. doi: 10.3390/ani5010138

Jones, E. (2022a) Owners put their horses first as economic situation bites, *Horse & Hound* 22 December, p. 10. https://www.horseandhound.co.uk/news/owners-put-their-horses-first-as-economic-situation-bites-812621 (Accessed 1 July 2023).

Jones, E. (2022b) New spotlight falls on ethics surrounding the use of horses, *Horse & Hound* 26 May, p. 4. https://www.scribd.com/article/576367484/New-Spotlight-Falls-On-Ethics-Surrounding-Use-Of-Horses (Accessed 1 July 2023).

Jones, E. (2023) Riding at risk as schools close but participation slightly up, *Horse & Hound*, 9 March p. 4. https://www.horseandhound.co.uk/news/250-riding-schools-close-in-four-years-818389 (Accessed 1 July 2023).

Jones McVey, R. (2018) *Reasonable Creatures: British Equestrianism and Epistemological Responsibility in Late Modernity*. University of Cambridge PhD Thesis.

Kanne, K. (2022) Riding, ruling, resistance: Equestrianism and political authority in the Hungarian Bronze Age, *Cultural Anthropology* 63, 289–329. doi:10.1086/720271

Katan, L. (2022) A view from the ground up, presentation to the National Equine Forum 3 March 2022. https://www.nationalequineforum.com/forum-2022/ (Accessed 22 March 2023).

Kelekna, P. (2009) *The Horse in Human History*. New York: Cambridge University Press.

Kennedy, M. (2021) An end to 'encouragement'- it's time for horseracing to follow the science about whipping, *RSPCA Blog*, https://www.rspca.org.uk/-/blog-time-for-horseracing-to-follow-science-about-whipping (Accessed 1 July 2023).

Kennedy, M. (2022) It's time for horse racing to wave goodbye to the whip, *RSPCA Blog*, 8th April 2022, https://www.rspca.org.uk/-/blog-time-for-horse-racing-to-wave-goodbye-to-the-whip (Accessed 1 July 2023).

Kinney, A., Eakman, A., Lassell, R. and Wood, W. (2019) Equine-assisted interventions for veterans with service-related health conditions: a systematic mapping review. *Military Medical Research* 6(1): 28 doi: 10.1186/s40779-019-0217-6

Kirkup, M. (2016) *Pit Ponies*. Newcastle upon Tyne: Summerhill Books.

Klecel, W. and Martyniuk, E. (2021) From the Eurasian steppes to the Roman circuses: A review of early development of horse breeding and management, *Animals* 11, 1859. doi:10.3390/ani11071859

Lambert, D. (2015) Master – Horse – Slave: Mobility, Race and Power in the British West Indies, c.1780–1838, *Slavery & Abolition* 36, 618–641 doi.org/10.1080/0144039X.2015.1025487

Landry, D. (2009) *Noble Brutes: How Eastern Horses Transformed English Culture*. Baltimore: Johns Hopkins Press.

Langdon, J. (1986) *Horses, Oxen and Technological Innovation: The Use of Draught Animals in English Farming 1066-1500*. Cambridge: Cambridge University Press.

Lash, S. and Urry, J. (1994) *Economies of Signs and Spaces*. London: Sage.

Latour, B. (1993) *We Have Never Been Modern*. Hemel Hempstead: Harvester Wheatsheaf.

Lawson, M. (2007) *The Battle of Hastings 1066*. Stroud: The History Press

League Against Cruel Sports (2021) Horse racing, https://www.league.org.uk/what-we-do/protect-animals/horse-racing/ (Accessed 1 July 2023).

Leaman, J. (1998) *Rural Opinion: Flushing Out the Truth*. Presentation to the Countryside Protection Group Conference, London, 24 February, London: MORI.

Lerner, H. and Silfverberg, G. (2019) Martha Nussbaum's capability approach and equine assisted therapy: an analysis for both humans and horses, pp.57–68 in J. Bornemark, P. Andersson and U. von Essen, (eds.) *Equine Cultures in Transition: Ethical Questions*. London: Routledge.

Levine, P. (2013). *The British Empire: Sunrise to Sunset*. (Second Edition). London: Routledge.

Librado, P. et al. (2021) The origins and spread of domestic horses from the Western Eurasian steppes, *Nature* 598, 634–40. doi:10.1038/s41586-021-04018-9

Löffelmann, T., Snoeck, C., Richards, J., Johnson, L., Claeys, P. and Montgomery, J. (2023) Sr analysis from only known Scandinavian cremation cemetery in Britain illuminate early Viking journey with horse and dog across the North Sea, *PLoS ONE* 18(2): e0280589. doi:10.1371/journal.pone.0280589

Lorimer, J. (2020) *The Probiotic Planet: Using Life to Manage Life*. Minneapolis: University of Minnesota Press.

Low Pay Commission (2020) *National Minimum Wage*. London: Low Pay Commission.

Lowe, P., Murdoch, J. and Cox, G. (1995a) A civilised retreat? Anti-urbanism, rurality and the making of an Anglo-centric culture, pp.63–82 in P. Healey, S. Cameron, S. Davoudi, S. Graham and A. Madani-Pour (eds) *Managing Cities: The New Urban Context*. Chichester: John Wiley and Sons.

Lowe, P., Murdoch, J. and Ward, N. (1995b) Beyond endogenous and exogenous models: Networks in rural development, pp.87–105 in J.D. van der Ploeg and G. van Dijk (eds.) *Beyond Modernization: The Impact of Endogenous Rural Development*, Assen, Netherlands: Van Gorcum.

Lowe, P., Ward, N., Ward, S. and Murdoch, J. (1995c) *Countryside Prospects, 1995-2010: Some Future Trends*, CRE, University of Newcastle

Marchand, W., Anderson, S., Smith, J., Hoopes, K. and Carlson, J. (2021) Equine-assisted activities and therapies for veterans with posttraumatic stress disorder: Current state, challenges and future directions, *Chronic Stress* 5, 1–11. doi:10.1177/2470547021991556

Marks, H. and Britton, D. (1989) *A Hundred Years of British Food and Farming – A Statistical Survey*. London: Taylor and Francis.

Marsden, T., Murdoch, J., Lowe, P., Munton, R. and Flynn, A. (1993) *Constructing the Countryside*, London: University College London Press.

Massey, D. (1984) *Spatial Divisions of Labour: Social Structures and the Geography of Production*. London: Macmillan.

May, A. (2013) *The Fox-Hunting Controversy, 1781-2004: Class and Cruelty*. Farnham: Ashgate.

McGivney, B., Han, H., Corduff, L., Katz, L., Tozaki, T., MacHugh, D. & Hill, E. (2020) Genomic inbreeding trends, influential sire lines and selection in the global Thoroughbred horse population, *Nature Scientific Reports* 10, 466. doi:10.1038/s41598-019-57389-5

McManus, P., Albrecht, G. and Graham, R. (2013) *The Global Horseracing Industry: Social, Economic, Environmental and Ethical Perspectives*. London: Routledge.

McShane, C. and Tarr, J. (2007) *The Horse in the City: Living Machines in the Nineteenth Century*. Baltimore: Johns Hopkins Press.

McWilliams, F. (2019) *Equine Machines: Horses and Tractors on British Farms c.1920-1970*. King's College London PhD Thesis, Department of History. https://kclpure.kcl.ac.uk/portal/en/theses/equine-machines(4c408217-1798-47ed-9acd-a55cbd7ca37d).html (Accessed 1 July 2023).

Meads, J. (1997) *Even Further Abroad – Nag, Nag, Nag*, BBC Two. https://www.youtube.com/watch?v=n5uSpsWOq2Q&t=35s (Accessed 6 July 2023).

Mellor, D. and Reid, C. (1994) Concepts of animal well-being and predicting the impact of procedures on experimental animals. pp.3–18 in Wellbeing International (eds.) *Improving the Well-Being of Animals in the Research Environment*. Glen Osmond, SA, Australia: Australian and New Zealand Council for the Care of Animals in Research and Teaching (ANZCCART).

Mendelson, S. and Crawford, P. (1998) *Women in Early Modern England, 1550-1720*. Oxford: Clarendon.

Ministry of Agriculture, Fisheries and Food (2000) *Current and Prospective Economic Situation in Agriculture*. Working Paper. London: MAFF.

Mitchell, P. (2015) *Horse Nations: The Worldwide Impact of the Horse on Indigenous Societies Post-1492*. Oxford: Oxford University Press.

Monroe, M., Whitworth, J., Wharton, T. and Turner, J. (2019) Effects of an equine-assisted therapy programme for military veterans with self-reported PTSC, *Society and Animals* 29, 1–14. doi:10.1163/15685306-12341572

Moore-Colyer, R. (2000) Aspects of the trade in British pedigree draught horses with the United States and Canada, c.1850–1920, *Agricultural History Review* 48, 42–59. https://www.jstor.org/stable/40275381

Morrison, M. (2007) Health benefits of animal-assisted interventions, *Complementary Health Practice Review* 12, 51–62. doi:10.1177/1533210107302397

Mortimer, I. (2023) *Medieval Horizons: Why the Middle Ages Matter*. London: Bodley Head

Murdoch, J. (2006) *Post-Structuralist Geography: A Guide to Relational Space*. London: Sage.

Murdoch, J. and Marsden, T. (1994) *Reconstituting Rurality: Class, Community and Power in the Development Process*. London: UCL Press.

Murphy, V. (2021) Queen Elizabeth has encyclopaedic knowledge of horse racing and it's her passion in life, Camilla says, *Town & Country Magazine*, 15 June. https://www.townandcountrymag.com/society/tradition/a36729442/camilla-quote-queen-elizabeth-love-horse-racing/ (Accessed 1 July 2023).

Murray, K. (1955) *Agriculture - History of the Second World War Series*. London: HMSO, p. 274

Murray, R. (2022) Support introduced to help sports cope with rising costs, *Horse & Hound* 16 June, p. 6. https://www.horseandhound.co.uk/news/horse-sport-takes-steps-to-support-competitors-during-cost-of-living-crisis-791454 (Accessed 1 July 2023).

Murray, R., Dyson, S., Tranquille, C. and Adams, V. (2006) Association of type of sport and performance level with anatomical site of orthopaedic injury diagnosis, *Equine Exercise Physiology* 7, *Equine Veterinary Journal Supplement* 36, 411–16. doi:10.1111/j.2042-3306.2006.tb05578.x

Nash, C. (2020a) Breed wealth: Origins, encounter value and the international love of breed, *Transactions of the Institute of British Geographers* 45, 849–61 doi:10.1111/tran.12383

Nash, C. (2020b) Kinship of different kinds: Horses and people in Iceland, *Humanimalia*, 12(1) 118–144. doi:10.52537/humanimalia.9426

National Farmers' Union (1999) *Routes to Prosperity for UK Agriculture*. London: National Farmers' Union.

Nosworthy, C. (2013) *A Geography of Horse-Riding: The Spacing of Affect, Emotion and (Dis)ability Identity through Human-Horse Encounters*. Newcastle upon Tyne: Cambridge Scholars Publishing.

Oliver, C. (2021) *Vegansim, Archives and Animals: Beyond Human Geographies*. London: Routledge.

Otter, C. (2011) Hippophagy in the UK: a failed dietary revolution, *Endeavour* 35, 80–90. doi:10.1016/j.endeavour.2011.06.005

Outram, A., Steer, N., Bendrey, R., Olsen, S., Kasparov, A., Zaibert, V., Thorpe, N. and Evershed, R. (2009) The earliest horse harnessing and milking, *Science* 323 (5919), 1332–35 doi:10.1126/science.ll68594

Overton, M. (1996) *Agricultural Revolution in England: The Transformation of the Agrarian Economy 1500–1850*. Cambridge University Press

Owers, R. and Chubbock, S. (2013) Fight the fat! *Equine Veterinary Journal* 45, 5, doi:10.1111/evj.12008

Paxman, J. (2021) *Black Gold: The History of How Coal Made Britain*. London: William Collins.

Percival, R. (2022) *The Meat Paradox: Eating, Empathy and the Future of Meat*. London: Little Brown.

Phelps, A., Gregory, R., Miller, R. and Wild, C. (2018) *The Textile Mills of Lancashire: The Legacy*. Lancaster: Oxford Archaeology North. https://historicengland.org.uk/images-books/publications/textile-mills-lancashire-legacy/textile-mills-lancashire-legacy/ (Accessed 1 July 2023).

Philo, C. and Wilbert, C. (2000) *Animal Spaces, Beastly Places: New Geographies of Human-Animal Relations*. London: Routledge.

Philo, C. and Wolch, J. (1998) Through the geographical looking glass: Space, place and society-animal relations. *Society & Animals* 6, 103–18. doi:10.1163/156853098x00096

Pickerill, J. (2005) '95% of thoroughbreds linked to one superstud', *New Scientist* 6 September, https://www.newscientist.com/article/dn7941-95-of-thoroughbreds-linked-to-one-superstud (Accessed 1 July 2023)

Piggott, S. (1981) Early prehistory, pp.3–62 in S. Piggott (ed.) *The Agrarian History of England and Wales Volume I-I Prehistory*. Cambridge: Cambridge University Press.

Pitt, M. (ed.) (2016) *The Queen and Her Horses*. London: Horse & Hound.

Poncet, P.A., Bachmann, I., Burger, D., Ceppi, A., Friedli, K., Klopfenstein, S., Maiatsky, M., Rieder, S., Rubli, S., Rüegg, P. and Trolliet, C. (2011) Considerations on Ethics and the Horse - Ethical Input for Ensuring Better Protection of the Dignity and Wellbeing of Horses. pp.64–67 in European States Studs Association (eds.) *Heritage Symposium of the European State Studs Association*, Lipica National Stud, October 13th, 2011. http://www.europeanstatestuds.org/de/termine/details/publikation-essa-kulturerbe-symposium.html?file=files/essa/archive/essaEuropeanStudCulture.pdf (Accessed 3 June 2022)

Poncet, P.A., Bachmann, I., Burkhardt, R., Ehrbar, B., Herrmann, R., Friedli, K., Leuenberger, H., Lüth, A., Montavon, S., Pfammatter, M. and Trolliet, C. (2022): *Ethical Reflections on the Dignity and Welfare of Horses and Other Equids – Pathways to Enhanced Protection*. Summary report. Bern: Swiss Horse Industry Council and Administration.

Porter, J. (1989) The development of rural society, pp.836–937 in G. Mingay (ed.) *The Agrarian History of England and Wales Volume VI 1750-1850 Part II*. Cambridge: Cambridge University Press.

PwC (2018) *The Contribution of Thoroughbred Breeding to the UK Economy and Factors Impacting the Industry's Supply Chain*. Newmarket: Thoroughbred Breeders Association.https://www.thetba.co.uk/wp-content/uploads/2018/09/TBA-Economic-Impact-Study-2018.pdf (Accessed 18 March 2022).

Raulff, U. (2018) *Farewell to the Horse: The Final Century of Our Relationship*. London: Penguin.

Riches, N. (1967) *The Agricultural Revolution in Norfolk*. (Second Edition). London: Frank Cass.

Ritvo, H. (2002) History and animal studies, *Society & Animals* 10, 403–06. doi:10.1163/156853002320936872

Rogers, S. (ed.) (2018) *Equine Behaviour in Mind: Applying Behavioural Science to the Way we Keep, Work and Care for Horses*. Sheffield: 5m Publishing

Rosen, B. and Zuckermann, W. (1982) *The Mews of London: A Guide to the Hidden Byways of London's Past*. Exeter: Webb and Bower.

Roth, M. (1998) Mounted police forces: a comparative history, *Policing: An International Journal of Police Strategies and Management* 21, 707–19. doi:10.1108/13639519810241700

Ryder, M. (1981) Livestock, pp.301–410 in S. Piggott (ed.) *The Agrarian History of England and Wales*. Cambridge: Cambridge University Press.

Schuurman, N. and Franklin, A. (2018) A good time to die: Horse retirement yards as shared spaces of interspecies care and accomplishment, *Journal of Rural Studies* 57, 110–117. doi:10.1016/j.jrurstud.2017.12.001

Shoard, M. (1980) *The Theft of the Countryside*. London: Temple Smith.

Sidnell, P. (2006) *Warhorse: Cavalry in Ancient Warfare*. London: Continuum.

Sinclair-Williams, K. and Sinclair-Williams, M. (2015) *Health and Safety in Horse Riding Establishments and Livery Yards*. London: Chartered Institute of

Environmental Health. https://www.cieh.org/media/1247/health-and-safety-in-horse-riding-establishments-and-livery-yards-what-you-should-know.pdf (Accessed 1 July 2023)

Singleton, J. (1993) Britain's military use of horses 1914-18, *Past & Present* 139, 178–204 doi:10.1093/past/139.1.178

Sleeman, P. (2023) *Bricks, Stones & Straw: Working Horses in Liverpool*. Stroud: Amberley Publishing.

SQW (2014) *Newmarket's Equine Cluster: The Economic Impact of the Horseracing Industry Centred upon Newmarket*. Cambridge: SQW https://www.westsuffolk.gov.uk/planning/Planning_Policies/local_plans/upload/D20-The-economic-impact-of-the-horseracing-industry-centred-upon-Newmarket.pdf (Accessed 1 July 2023).

Stott, C. (2009). *Crowd Psychology and Public Order Policing: An Overview of Scientific Theory and Evidence*. Report submitted to the HMIC inquiry into the policing of the London G20 protests. Liverpool: University of Liverpool School of Psychology. https://www.workingwithcrowds.com/wp-content/uploads/2018/05/Dr_Clifford_Scott_Crowd_Psychology_and_Public_Order_Policing.pdf (Accessed 1 July 2023).

Sutherland, L.A. (2021) Horsification: Embodied gentrification in rural landscapes, *Geoforum* 126, 37–47. doi:10.1016/j.geoforum.2021.07.020

Swart, S. (2007) But where's the bloody horse?: Textuality and corporeality in the "animal turn", *Journal of Literary Studies* 23, 271–292, doi:10.1080/02564710701568121

Swart, S. (2010) *Riding High: Horses, Humans and History in South Africa*. Johannesburg: University of Witwatersrand.

Symanski, R. (1994) Contested realities: Feral horses in Outback Australia, *Annuls of the Association of American Geographers* 84, 251–69. doi:10.1111/j.1467-8306.1994.tb01738.x

Taylor, J. (2022) *'I Can't Watch Anymore': The Case for Dropping Equestrian from the Olympic Games*. Copenhagen: Epona Publishing.

Taylor, M. (2007) Hunt supporters claim legislation has backfired, *The Guardian* 27 December 2007, p. 7.

Taylor, W. and Barrón-Ortiz, C. (2021) Rethinking the evidence for early horse domestication at Botai, *Nature* (Scientific reports) 11, 7740. doi:10.1038/s41598-021-86832-9

The Henley Centre (2004) *A Report of Research on the Horse Industry in Great Britain*. Report to Defra and the British Horse Industry Confederation. London: Defra.

Thirsk, J. (1967) Farming techniques, pp.161–255 in J. Thirsk (ed.) *The Agrarian History of England and Wales Volume IV 1500-1640*. Cambridge: Cambridge University Press.

Thirsk, J. (1984) *The Rural Economy of England: Collected Essays*. London: Hambledon Press.

Thomas, B. (2021) The truth of horse evolution – Part 2, Youtube https://youtu.be/9vYcTSyf8bk (Accessed 1 July 2023).

Thomas, H. (1997) *The Slave Trade: The Story of the Atlantic Slave Trade: 1440–1870*. London: Simon & Schuster.

Thomas, K. (1983) *Man and the Natural World: Changing Attitudes in England, 1500-1800*. London: Allen Lane.

Thompson, C. (2008) *Harnessed: Colliery Horses in Wales*. Cardiff: National Museum of Wales.

Thompson, F.M.L. (1970) *Victorian England: The Horse-Drawn Society*; London: Bedford College.

Thompson, F.M.L. (1976) Nineteenth century horse sense, *Economic History Review* 29, 60–81. doi:10.2307/2594507

Thompson, F.M.L. (ed.) (1983a) *Horses in European Economic History: A Preliminary Canter*. Reading: British Agricultural History Society.

Thompson, F.M.L. (1983b) Horses and hay in Britain 1830 to 1900, pp.50–72 in F.M.L. Thompson (ed.) *Horses in European Economic History: A Preliminary Canter*. Reading: British Agricultural History Society.

Thoroughbred Breeders' Association (2023) Third thoroughbred breeding industry economic impact study provides blueprint for future progress, *Press Release* 18th January https://www.thetba.co.uk/resource/third-thoroughbred-breeding-industry-economic-impact-study-provides-blueprint-for-future-progress.html (Accessed 1 July 2023).

Thrift, N. (2004) Summoning life, pp.81–103 in P. Cloke, P. Crang and M. Goodwin (eds.) *Envisioning Human Geographies*. London: Arnold.

Thrift, N. (2021) *Killer Cities*. London: Sage.

Tomlinson, R. (1986) A geography of flat-racing in Great Britain, *Geography* 7 1, 228–39. https://www.jstor.org/stable/40571124

Townsend, M. (2008) Hunting ban sparks a rural boom, *The Observer* 24 February 2008, p. 26

Turner, R. (2022) How all equestrians can help make the world a greener place, *Horse and Hound* 4 January https://www.horseandhound.co.uk/news/how-all-equestrians-can-help-make-the-world-a-greener-place-771531 (Accessed 1 July 2023).

Two Circles (2015) *The National Equestrian Survey. 2015. Overview Report*. Wakefield: British Equestrian Trade Association.

UK Government (2021) *Net Zero Strategy: Build Back Greener*. London: Stationary Office. https://www.gov.uk/government/publications/net-zero-strategy (Accessed 1 July 2023)

UK Government (2023) *UK Greenhouse Gas Inventory, 1990-2021*. London: Department for Energy Security and Net Zero. Annexes. https://unfccc.int/documents/627789 (Accessed 1 July 2023)

Uldahl, M. and Clayton, H. (2019) Lesions associated with the use of bits, nosebands, spurs and whips in Danish competition horses, *Equine Veterinary Journal* 51 154–162. doi:10.1111/evj.12827

Urry, J. (1995) *Consuming Places*. London: Routledge.

Urry, J. (2007) *Mobilities*. Cambridge: Polity Press.

Waddington, D. (1992) *Contemporary Issues in Public Disorder: A Comparative and Historical Approach*. London: Routledge.

Waddington, D. (2017) The battle of Orgreave: Afterword, pp.366–74 in R. Page (ed.) *Protest: Stories of Resistance*. Manchester: Comma Press.

Wade, C., Guilotto, S., Sigurdsson, E., Zoli, M., Gnerre, S., Imsland, F., Lear, T., Adelson, D., Bailey, E., Bellone, R., Blocker, H., Distl, O., Edgar, R., Garber, M., Leeb, T., Mauceli, E., MacLeod, J., Penedo, M., Raison, J., Sharpe, T., Vogel, J., Andersson, L., Antczak, D., Biagi, T., Biins, M., Chowdhary, B., Coleman, S., Della Valle, G., Fryc, S., Guerin, G., Hasegawa, T., Hill, E., Jurka, J., Kiialainen, G., Lindgren, A., Liu, J., Magnani, E., Mickelson, J., Murray, J., Nergadze, S., Onofrio, R., Pedroni, S., Piras, M., Raudsepp, T., Rocchi, M., Roed, K., Ryder, O., Searle, S., Skow, L., Swinburne, J., Syvanen, A., Tozaki, T., Valberg, S., Vaudin, M., White, J., Zody, M., Lander, E. and Lindblad-Toh, K. (2009) Genome sequence, comparative analysis, and population genetics of the domestic horse. *Science* 326 (5954): 865–67. doi:10.1126/science.1178158

Wadham, H. (2020) Horse matters: re-examining sustainability through human-domestic animal relationships, *Sociologia Ruralis* 60, 530–50. doi:10.1111/sori.12293

Wadham, H., Wallace, C. and Furtado, T. (2023) Agents of sustainability: How horses and people co-create, enact and embed the good life in rural places, *Sociologia Ruralis* 63, 390–414. doi:10.1111/soru.12387 (accessed 1 July 2023).

Walmsley, D. (2012) *Horseracing UK*. London: Mintel (Accessed Business and IP Centre, British Library, 24 March 2022).

Walmsley, D. (2019) *Special Interest Holidays UK*. London: Mintel. (Accessed Business and IP Centre, British Library, 24 March 2022).

Walmsley, D. (2021) *Mintel Gambling Review UK 2021*. London: Mintel. (Accessed Business and IP Centre, British Library, 24 March 2022).

Ward, N. (1999) Foxing the nation: the economic (in)significance of hunting with hounds in Britain, *Journal of Rural Studies* 15, 389–403. doi:10.1016/S0743-0167(99)00005-4

Ward, N. (2023) *Net Zero, Food and Farming: Climate Change and the UK Agri-Food System*. London: Routledge.

Ward, N. and Lowe, P. (2007) Blairite modernisation and countryside policy, *The Political Quarterly* 78 (3), 412–21. doi:10.1111/j.1467-923X.2007.00869.x

Warmouth, V. et al. (2012) Reconstructing the origin and spread of horse domestication in the Eurasian steppe, *Proceedings of the National Academy of Sciences* 109, 8202–06. doi:10.1073/pnas.1111122109

Westendorf, M., Williams, C., Murphy, S., Kenny, L. and Hashemi, M. (2019) Generation and management of manure from horses and other equids, pp.145–163 in H. Waldrip, P. Pagliari and Z. He (eds.) *Animal Manure: Production, Characteristics, Environmental Concerns and Management*. Madison: Wiley.

Whatmore, S. (2002) *Hybrid Geographies: Natures, Cultures, Spaces*. London: Sage.

Whittet, A. (ed.) (1986) *The Bridleways of Britain*. London: Whittet Books.

Wilkin, S., Ventresca Miller, A., Fernandes, R., Spengler, R., Taylor, W., Brown, D., Reich, D., Kennett, D., Culleton, B., Kunz, L., Fortes, C., Kitova, A., Kuznetsov, P., Epimakhov, A., Zaibert, V., Outram, A., Kitov, E., Khokhlov, A., Anthony, D. and Boivin, N. (2021) Dairying enabled Early Bronze Age Yamnaya steppe expansions, *Nature* 598, 629–633. doi:10.1038/s41586-021-03798-4

Williams, E. (2022) *Capitalism and Slavery*. London: Penguin (First published in 1944).

Wolch, J. and Emel, J. (1995) Bringing the animals back in. *Environment and Planning D: Society and Space* 13, 632–6. soi: 10.1068/d130632

Woodhouse, J. (2019) *The Horserace Betting Levy*. House of Commons Library Briefing Paper No. 7368. London: House of Commons https://commonslibrary.parliament.uk/research-briefings/cbp-7368/ (Accessed 1 July 2023).

Woods, A. (2011) From cruelty to welfare: the emergence of farm animal welfare in Britain, 1964-71, *Endeavour* 36, 14–22. doi:10.1016/j.endeavour.2011.10.003

World Horse Welfare (2022) Sector leaders discuss involvement of horses in sport, *Press Release* https://www.worldhorsewelfare.org/news/sector-leaders-discuss-involvement-of-horses-in-sport (Accessed 1 July 2023)

World Horse Welfare and EuroGroup4Animals (2015) *Removing the Blinkers: The Health and Welfare of European Equidae in 2015*. Snetterton: World Horse Welfare. https://www.eurogroupforanimals.org/files/eurogroupforanimals/2021-12/EU-Equine-Report-Removing-the-Blinkers_0.pdf (Accessed 1 July 2023).

Wright, H. (2017) *Outside Time: A Personal History of Prison Farming and Gardening*. Eightslate: Placewise Press.

Wrigley, E. (2006) The transition to an advanced organic economy: half a millennium of English agriculture, *Economic History Review* 59, 435–480. https://www.jstor.org/stable/3805970

Yarwood, R. and Evans, N. (1998) New places of 'old spots': The changing geographies of domestic livestock animals, *Society and Animals* 6, 137–65. doi:10.1163/156853098X00122

Zasada, I., Berges, R., Hilgendorf, J. and Piorr, A. (2013) Horsekeeping and the periurban development in the Berlin Metropolitan Region. *Journal of Land Use Science* 8, 199–214. doi:10.1080/1747423X.2011.628706

Zeuner, F. (1963) *A History of Domesticated Animals*, London: Hutchinson.

Zhu, X., Suarez-Jimenez, B., Zilcha-Mano, S., Lazarov, A., Arnon, S., Lowell, A., Bergman, M., Ryba, M., Hamilton, A. and Hamilton, J. (2021) Neural changes following equine-assisted therapy for posttraumatic stress disorder: A longitudinal multimodal imaging study, *Human Brain Mapping* 42, 1930–39 doi:10.1002/hbm.25360

Index

Pages in *italics* refer to figures and pages in **bold** refer to tables.

Printed in the United States
by Baker & Taylor Publisher Services